coloryourselfsmart
Geography

The fun, visual way to teach yourself about anything and everything

By Dan Cowling
Illustrations by Mark Franklin

THUNDER BAY
P·R·E·S·S

San Diego, California

Design by Shawn Dahl, dahlimama inc

 Thunder Bay Press
An imprint of the Baker & Taylor Publishing Group
10350 Barnes Canyon Road, San Diego, CA 92121
www.thunderbaybooks.com

ISBN-13: 978-1-60710-216-8
ISBN-10: 1-60710-216-1

Printed in China
1 2 3 4 5 15 14 13 12 11

CONTENTS

INTRODUCTION

"Without geography you're nowhere."
JIMMY BUFFET

GEOGRAPHY IS AN AMAZING SUBJECT, and over time it is possible that you may even come to love it. It is the only subject that helps us make sense of the world and what is going on around us. By its very nature, it is constantly evolving and changing as researchers around the world push the boundaries of the subject further, often overlapping with other disciplines in an attempt to make sense of some of the most pressing issues facing humanity—climate change, vast inequalities between nations and within nations, and the understanding of natural disasters (and how to prevent them from becoming human ones).

To the casual observer, geography may seem a disparate subject, perhaps lacking direction, or spread across too many disciplines. Books have been written trying to define the subject, but it seems that central to understanding geography is knowing that the subject relates to the "interaction between people and Earth." A large part of this book is focused on introducing this interaction to the reader, through, for example, understanding how volcanoes and earthquakes occur and what impacts they can have on people.

But geography can be more than this—many academic geographers are interpreting geography as a subject of "spatial analysis" and are therefore eager to explore spatial relationships across Earth. Collecting data on HIV infection rates, for example, and mapping these across the world gives a geographer the opportunity to see the patterns of infection that have emerged and puts a geographer in the powerful position to ask why the pattern is like it is and what the contributing factors are.

This book is divided into nine themes which reflect different aspects of geography. These themes are:

1. Getting to Know the World: Orient yourself with the key physical and human characteristics of our planet. Exploring the physical makeup of planet Earth was the founding of geography as a subject.

2. Getting to Know the United States: Explore the physical and human characteristics of the country in which you live and understand the history behind the geography of the United States.

3. Natural Processes: Gain an understanding of how our planet works and explore the key processes that help sustain life on Earth.

4. People: Learn about the impact that people have had on planet Earth, as well as exploring aspects of inequalities in wealth and lifestyle across the planet.

5. Resources and the Environment: Investigate where the major resources of Earth are found and how they are distributed. You will also learn about the impact that our use of natural resources is having on the environment.

6. Trade Groups and Organizations: **Gain an insight into how countries of the world cooperate with each other as they form part of the world economy.**

7. Emerging Superpowers: **Learn some of the key geographical features of the countries that are on the brink of becoming world superpowers—Brazil, Russia, India, and China (the BRICs).**

8. Areas of Conflict: **Gain an understanding of some of the conflicts that have occurred around the world in recent times and the impacts that they have had on both the countries involved and the people who live there.**

9. Technology: **Gain a brief insight into how geography is beginning to try and make sense of the "technological age" that is now upon us.**

Geography enables us to see beyond the patterns and processes, and when it is at its best, enables us to ask questions about what is happening on our planet. Perhaps most importantly, it puts forward suggestions to make things better in the future. Many of the coloring tasks in this book will enable you to understand physical processes and map human features. Please don't just stop when you have finished coloring, that is just the start of your geographical journey—start asking questions about who, what, why, when, and how? The coloring exercises start producing patterns that can be intriguing, and geography puts you in a powerful position to start thinking about what you are seeing.

NOTE FROM THE AUTHOR

Every effort has been taken to ensure that the facts in this book are as accurate as possible at the time of going to print. Naturally, given time, some of the measurements and numbers will change (such as the population of a given country). There are also many geographical "facts" which are disputed among geographers themselves (e.g. the length of a river can be difficult to calculate depending on where the location of the source, or mouth is identified). This is why you see so many different figures coming from varied sources both online and in print. To that end, please enjoy the facts that you learn along the way, but be aware that geography, like so many other disciplines, is often open to interpretation. Below I've listed a few of the key trusted references I used while writing this book should you want to read further:

AS Geography for Edexcel, Oxford University Press, 2008

A2 Geography for Edexcel, Oxford University Press, 2009

Atlas of the Real World, Thames & Hudson, 2008

GCSE Geography for Edexcel B, Oxford University Press, 2010

Longman Student Atlas in association with the Geographical Association, Pearson, 2005

The New Wider World, Nelson Thornes, 2003

New Key Geography, Nelson Thornes, 2006

Philip's Modern School Atlas in association with the Royal Geographical Society, 2006

HOW TO USE THIS BOOK

A pack of eight artists' studio-quality coloring pencils are included with this book, however, you should feel free to supplement these colors with your own as you see fit—the more colors you have available to you, the more enjoyable you will find memorizing the material. A colored plate section is included at the back of the book, showing all of the featured illustrations colored up for your reference, and a color key accompanies those illustrations that are more complicated in nature, for those readers who prefer a little more guidance.

Keep your pencil tips sharp with the accompanying sharpener and use the eraser to correct any mishaps. The eraser is more effective when used on lighter washes of color and may not be so helpful if you have pressed too hard with one of the darker colors.

WHAT IS A CHLOROPLETH MAP?

A chloropleth map is one in which areas are shaded or colored in proportion to the measurement being displayed. It is a commonly used tool by geographers to enable them to easily spot patterns and trends as it helps to visualize the data across a geographic area.

COLORED PENCIL TECHNIQUES

Numerous finishes can be achieved with colored pencils, depending on how they are used. Artists' quality pencils can be smudged for a watercolor effect, and different colors can be blended to create other colors. When blending pencil colors, it is best to lay down the lighter colors first and overlay the darker colors to achieve the desired effect.

Pressing harder or lighter on the paper will also give a different shade of color. It is easier to use the side of a pencil point to wash large areas with color and use a sharp point for those areas that require more careful coloring or more detail.

ABOUT THE *COLOR YOURSELF SMART* SERIES

Color Yourself Smart is a revolutionary new series designed to help improve your memory and make learning easy. Leading memory and learning experts agree that color and illustrations help reinforce difficult subject matter and greatly increase your chances of both creating visual memories and recalling that information faster. So if you find it difficult to remember something—even after you've just read it—then it's time to start coloring your way to faster learning and a sharper memory!

Continents and Oceans

EARTH IS THE THIRD PLANET FROM THE SUN and the fifth-largest in the solar system. It is believed to be around 4.55 billion years old, although some scientists disagree on this. The oldest rocks that have been found on Earth date to around 3.9 billion years ago, and there is an agreement that Earth must be at least as old as these rocks.

The continents of the world have not always been in their current positions. In 1915, Alfred Wegener, a German geologist and meteorologist, first put forward the theory of "continental drift." He stated that the parts of Earth's crust float slowly over a liquid core and over time the continents of the world move and shift. He also argued that the surface of Earth was once covered by a giant super-continent that he named *Pangea* (meaning "All-Earth"). This giant super-continent slowly broke up over time forming the seven continents that make up the Earth that we know today: North America, South America, Europe, Asia, Africa, Australia and Oceania, and Antarctica.

More than two-thirds of the surface of Earth is covered by seas and oceans and these contain over 97 percent of the world's water. The oceans of Earth are unique in the solar system as no other planet has liquid water (although recent finds on Mars indicate that it may have once had some liquid water in the recent past). Life on Earth originated in the seas, and the oceans continue to be home to an incredibly diverse range of life. The depth of the seas and oceans varies considerably with many areas of land having fairly shallow water close to them where continental shelves extend out from the coast. The oceans do not remain still—ocean currents have a major effect on the world's climate patterns as they carry cold or warm water around the globe.

10 THINGS TO REMEMBER

1. **Earth:** Despite being called "Earth," only 29 percent of the surface is actually "earth." The other 71 percent of the surface is made up of water.

2. **Earth's orbit:** It takes 365¼ days for the Earth to orbit the sun. We have a leap year every four years when an extra day is added to make up the four quarter days.

3. **The speed of Earth:** Earth travels through space at 66,700 mph (107,343 km/h).

4. **The tilt of Earth:** Earth has a tilt of about 23½ degrees on its axis. It is this tilt that causes the seasons.

5. **Surface area:** The surface of Earth is around 197 million sq miles (510 million km^2).

6. **Water:** Earth is the only planet in the solar system to have water in its three states of matter: as a solid (ice), as a liquid (oceans and seas), and as a gas (clouds).

7. **The power of the sun:** Sunlight can penetrate ocean water to a depth of 240 feet (73 m).

8. **Fresh water:** Fresh water makes up only 3 percent of Earth's water on the earth. Of this 3 percent, over 2 percent is frozen in ice sheets and glaciers, meaning that less than 1 percent fresh water is found in lakes, rivers, and underground.

9. **The oceans:** On average, Earth's oceans are only 2 miles (3.2 km) deep.

10. **Life on Earth:** Although Earth is around 4.55 billion years old, life (as we know it) has only existed in the last 150 to 200 million years.

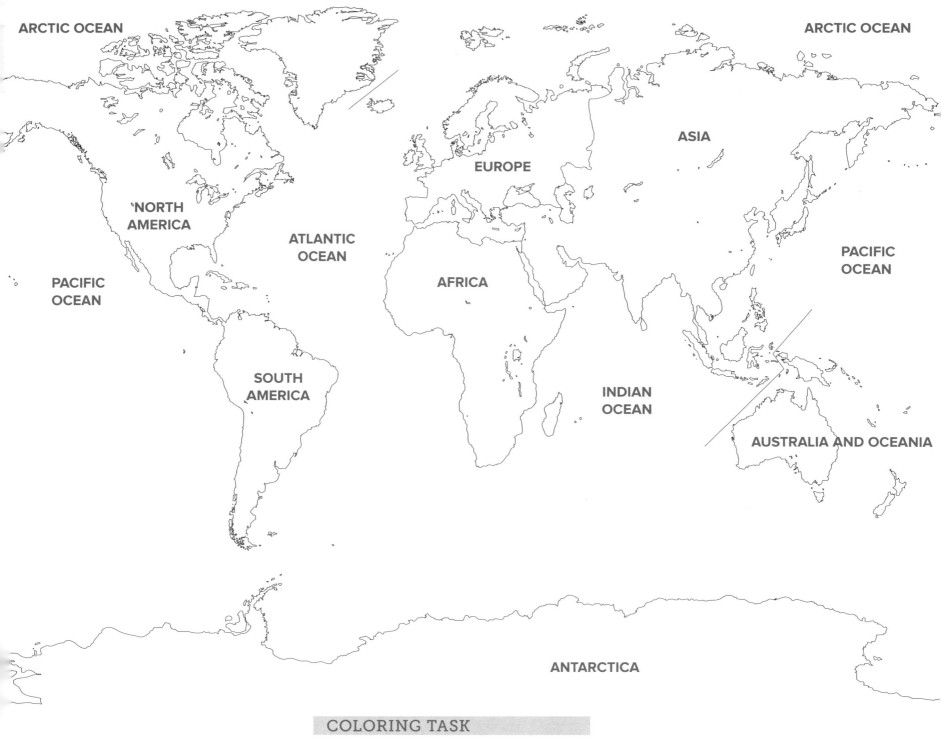

ARCTIC OCEAN

ARCTIC OCEAN

ASIA

EUROPE

NORTH
AMERICA

ATLANTIC
OCEAN

PACIFIC
OCEAN

PACIFIC
OCEAN

AFRICA

SOUTH
AMERICA

INDIAN
OCEAN

AUSTRALIA AND OCEANIA

ANTARCTICA

COLORING TASK

Color each of the continents a separate
color and the five oceans in blue.

Political—Countries of the World

THERE ARE ALMOST 200 SEPARATE COUNTRIES in the world today. Many national borders are influenced by physical features such as mountains, lakes, and rivers, although other human factors such as language, ethnicity, culture, or religion can be dividing barriers. In many former colonial lands, particularly in Africa, straight-line boundaries indicate how lands were carved up by drawing lines on maps. While many countries have well-established borders that have been present for many centuries, others are still disputing theirs.

How you define the exact number of countries is open to interpretation, and the number can vary depending on which source you use. The United Nations has 192 member states—but this does not include all the countries in the world as there are two officially recognized independent states that are not members of the UN—The Vatican City and Kosovo. This makes 194 countries.

The U.S. State Department officially recognizes the number of countries in the World at 194 (the 192 members of the UN plus The Vatican City and Kosovo), but it fails to recognize Taiwan as an independent state, separate from China.

There are also a number of territories that could be considered "countries," but they do not have the status of independent states. These include: Palestine, Lapland, Greenland, Bermuda, and Puerto Rico. It is also worth noting that parts of the United Kingdom—England, Scotland, Wales, and Northern Ireland—are often mistakenly referred to as separate countries—they are actually consituent parts of the UK.

10 THINGS TO REMEMBER

1. **Country with the highest proportion of female prisoners:** Maldives, 22 percent

2. **Country with the shortest maternity leave:** The United Arab Emirates (UAE), with just 45 days' maternity leave.

3. **The worst place to go to school:** In Nigeria, there are 8.2 million children out of school—more than any other country in the world!

4. **The best university in the world:** Harvard University in the United States, with an overall score of 96.1 percent in the *Times Higher Education World University Rankings* in 2010.

5. **Country with the most paid vacation time:** Finland, with 44 days off per year (including vacation and public holidays).

6. **Country with the highest density of millionaires:** Singapore, with 11.4 percent of households having at least one millionaire.

7. **Country with the highest per capita wine consumption:** The Vatican City—probably due to the regular administration of the Eucharist (or Holy Communion).

8. **Country with the highest suicide rate:** South Korea, with a suicide rate of around 22 per 100,000 people—probably partly due to high levels of depression as part of the push for achievement.

9. **Country with the most expensive cell phone plan:** Canadians pay the highest for a complete package of voice, text, and data, at around $67.50 per month.

10. **Country that Americans adopt from the most:** China, with 3,001 visas issued for adoption in 2009.

COLORING TASK

Color each of the countries named in the 10 Things to Remember, shown on the map in bold.

COLORING KEY

1. ☐ 2. ☐ 3. ☐ 4. ☐ 5. ☐ 6. ☐ 7. ☐ 8. ☐ 9. ☐ 10. ☐

Physical Geography

PHOTOGRAPHS TAKEN FROM SPACE show Earth as a blue planet with over 70 percent of the surface covered by water as found in the seas and oceans. Although the planet is circular like a ball, satellite images have shown that it is more like a slightly flattened ball, with a bulge at the equator and slightly flattened at the poles. We have divided Earth with two key imaginary lines: the equator and the prime meridian. The equator runs around Earth at the halfway point between the two poles and divides the globe into the Northern and Southern hemispheres. The prime meridian (often referred to as the Greenwich meridian, named after Greenwich in London, UK) runs through 0° and 180° longitude creating the Eastern and Western hemispheres.

Earth can also be divided into land and water hemispheres when viewed from space. The Pacific Ocean dominates the water hemisphere, giving a view that is completely made up of water since the Pacific stretches halfway across Earth at its widest point.

Earth's mountains make up about one-fifth of the surface of the land, occur in 75 percent of countries, and provide over 80 percent of our fresh water. The highest mountain is Mount Everest at around 29,035 feet (8,850 m) tall, located in the Himalayas with 14 of the next highest mountains in the world for company! The Andes Mountains in South America form the longest mountain range in the world, stretching over 4,500 miles (7,242 km) from north to south. Mountains can form in several different ways, but the most common is "fold mountains." These occur when two of Earth's plates collide head on, and their edges crumble as they push together. Large areas of rock are thrust upward, and over millions of years these form into mountains. The Himalayan Mountains were formed in this way when India crashed into Asia and pushed up the tallest mountain range.

10 THINGS TO REMEMBER

1. **Highest point:** The highest mountain in the world is Mount Everest (China/Nepal), standing at 29,035 feet (8,850 m) above sea level.

2. **Lowest point on a land-based continent:** The lowest point is the Dead Sea (in the Middle East), at 1,368 feet (417 m) below sea level.

3. **Largest ocean:** The largest ocean is the Pacific Ocean, at 64,186,008 sq miles (166,241,000 km²).

4. **Deepest ocean point:** The deepest point in the ocean is Challenger Deep (Pacific Ocean), at 36,745 feet (11,200 m) below sea level.

5. **Largest lake:** The largest lake is the Caspian Sea (in Asia), at 143,243 sq miles (371,000 km²).

6. **The equator:** The equator is an imaginary line that runs around the middle of Earth and has a circumference of 24,901 miles (40,075 km).

7. **Longest river:** The longest river is the River Nile (Africa), at 4,240 miles (6,825 km).

8. **From pole to pole:** The diameter from the North Pole to the South Pole is 7,900 miles (12,714 km).

9. **Largest island:** The largest island is Greenland, at 324,327 sq miles (840,004 km²).

10. **The age of Earth:** Earth is believed to be around 4.5 billion years old.

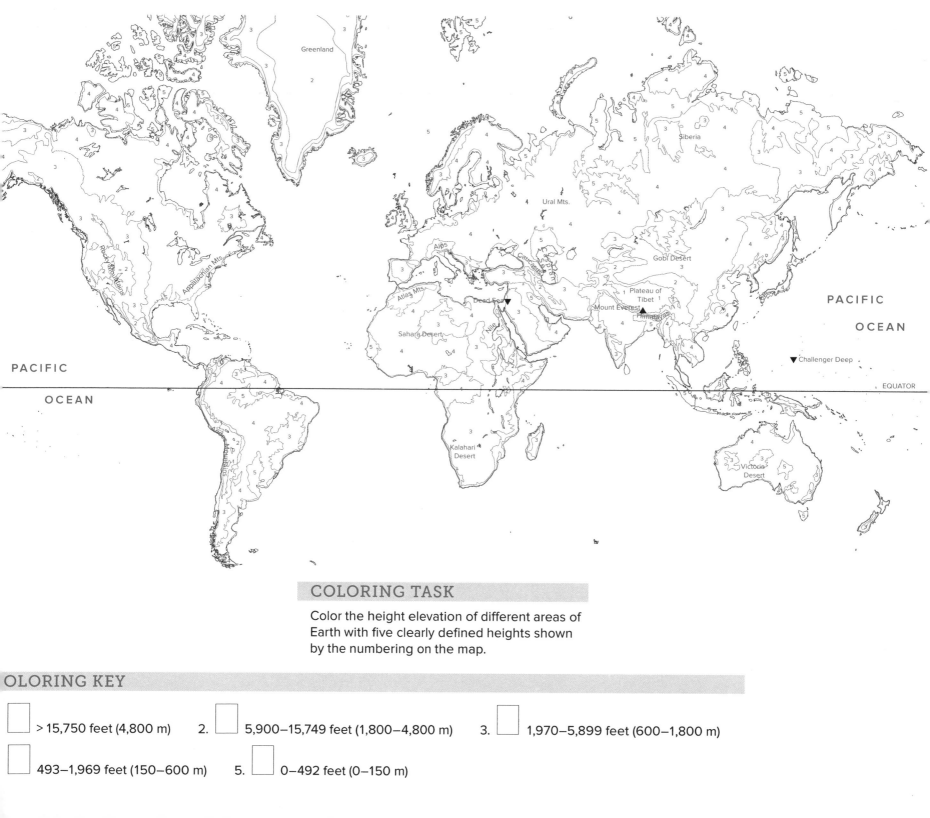

COLORING TASK

Color the height elevation of different areas of
Earth with five clearly defined heights shown
by the numbering on the map.

OLORING KEY

| | > 15,750 feet (4,800 m) | 2. | | 5,900–15,749 feet (1,800–4,800 m) | 3. | | 1,970–5,899 feet (600–1,800 m) |

| | 493–1,969 feet (150–600 m) | 5. | | 0–492 feet (0–150 m) |

North America—Political and Physical

NORTH AMERICA IS THE WORLD'S THIRD-LARGEST CONTINENT, covering an area of around 9.5 million sq miles (24.6 million km^2)—about 5 percent of Earth's surface. It extends from a narrow base in the tropics to progressively wider portions in middle latitudes and the Arctic polar margins. It can be divided into several distinctive physical regions; the mountains of the Western Cordillera in the West move to the Great Plains of central Canada and the U.S. which drain by the Mississippi River into the Gulf of Mexico.

The two largest countries of North America are Canada and the U.S., and these are the second- and third-largest countries in the world respectively. The continent also includes Greenland and the countries of Central America and the Caribbean.

The only land connection to North America is through South America at the narrow Isthmus of Panama, and some argue that Panama is part of Central America as it sits separately on the Caribbean plate. However, Central America is considered to be too small to be a distinctive continent on its own, and Central America is considered to be a region of the continent of North America. Greenland, which is geographically a part of North America, is not considered to be part of the continent politically.

North America experiences a wide range of climate zones. In the north, Canada experiences mainly polar and subarctic climates. The U.S. has a warm, temperate climate on the eastern coast, changing to cool, continental inland. In the south, Central America and the Caribbean experience tropical, hot, and humid conditions.

10 THINGS TO REMEMBER

1. **How many countries are there in North America?** 24 countries make up the continent of North America.

2. **How many people live there?** About 528 million people live in North America.

3. **What languages are spoken there?** Most North Americans speak English, Spanish, or French.

4. **What is the largest city in North America?** Mexico City, Mexico, is the largest city in North America.

5. **How did North America become named?** North America was named after the Italian explorer Amerigo Vespucci.

6. **What is the biggest island on the planet?** The largest island on the planet is Greenland.

7. **What is the longest river in North America?** The longest river in North America is the Mississippi River, at 2,320 miles (3,733 km).

8. **What is the highest point in North America?** The highest point is Mount McKinley in Alaska, at 20,321 feet (6,194 m) above sea level.

9. **What is the lowest point in North America?** The lowest point is Death Valley in California, at 282 feet (86 m) below sea level.

10. **What is the largest lake in North America?** The largest lake is Lake Superior, at 31,700 sq miles (82,102 km²).

COLORING TASK

Color the countries of North America to match the color plate on page 114, and color the main physical features (such as rivers and mountains).

South America—Political and Physical

SOUTH AMERICA IS THE WORLD'S FOURTH-LARGEST CONTINENT, covering roughly 6.9 million sq miles (17.8 million km²). It stretches from 12° north of the equator to the island of Cape Horn, 56° south. It is located predominantly in the Southern Hemisphere. The Andes Mountain range and the Amazon River are the dominant physical features of the continent, although highland plateaus features such as the Mato Grosso have been formed from the remnants of older, eroded shield mountains.

Over half of the land area of South America is made up of the country of Brazil, although there are in total 12 countries making up the continent. Before European settlers arrived in South America, the Inca civilization was the dominant force there. In the 1500s, Spain and Portugal colonized much of South America. As a result, much of South America still speaks Spanish, and Portuguese is the primary language of Brazil. The population is spread unevenly across the continent—there are areas of high population density along the coasts and in the north, while much of the interior of the continent remains empty, or sparsely populated. After 50 years of high population growth, many countries in South America have seen their population growth slow dramatically in recent years.

Much of South America experiences a tropical climate. However, the west of the continent is much drier and some areas of semiarid and hot desert exist. The climate to the extreme south is tundralike where the Patagonia ice sheet and several glaciers are found.

10 THINGS TO REMEMBER

1. **How many countries are there in South America?** Twelve countries make up the continent of South America.

2. **How many people live there?** About 385 million people live in South America.

3. **What languages are spoken there?** Most South Americans speak Spanish or Portuguese.

4. **What is the largest city in South America?** São Paulo, Brazil, is the largest city in South America.

5. **What makes South America unique?** South America claims the highest waterfall in the world, Angel Falls in Venezuela.

6. **The country of Chile . . .** has the world's most southerly located community at Puerto Toro.

7. **What is the longest river in South America?** The longest river in South America is the Amazon River, at nearly 4,000 miles (6,430 km).

8. **What is the highest point?** The highest point is Cerro Aconcagua in Argentina, at 22,831 feet (6,959 m) above sea level.

9. **What is the lowest point?** The lowest point is Peninsula Valdés in Argentina, at 131 feet (40 m) below sea level.

10. **What is the largest lake?** The largest lake is Lake Titicaca on the border of Peru and Bolivia, at 3,232 sq miles (8,370 km²).

COLORING TASK

Color the countries of South America to match the color plate on page 115, and color the main physical features (such as rivers and mountains).

Caribbean Sea

ATLANTIC OCEAN

Barranquilla
Isthmus of Panama
Maracaibo
Valencia ★ Caracas
Medellin
Barquisimeto
VENEZUELA
Rio Orinoco
Georgetown
Paramaribo
Angel Falls ■ GUYANA
Cayenne
Rio Magdalena
★ Bogota
Rio Orinoco
SURINAME
FRENCH GUIANA
Cali
COLOMBIA
Marajo Island
★ Quito
ECUADOR
Rio Negro
Amazon R.
Belem
Guayaquil
Gulf of Guayaquil
Amazon
Amazon R.
Manaus
Fortaleza
Parinas Point
Amazon R.
Basin
Rio Madeira
Sao Goncalo
PERU
Andes Mountains
Rio Jurua
Rio Purus
Rio Tapajos
BRAZIL
Recife
Rio Xingu
Rio Teles Pires
Rio Tocantins
Lima
Rio Madre de Dios
Mato Grosso
Rio Sao Francisco
Salvador
Plateau
Lake Titicaca
BOLIVIA
Rio Araguaia
★ Brasilia
★ La Paz
Goiania
Sucre ★
Belo Horizonte
PARAGUAY
Rio Paraguay
Rio Parana
Campinas
Rio De Janeiro
CHILE
Atacama Desert
Sao Paulo
PACIFIC OCEAN
Curitiba
Asuncion ★
Rio Parana
Rio Uruguay
Porto Alegre
ARGENTINA
Andes Mountains
Cerro Aconcagua
Rosario
URUGUAY
★ Santiago
Pampas
Buenos Aires
Montevideo

ATLANTIC OCEAN

Gulf of San Matias
Valdes Penninsula
Isla Grande de Chiloe
Patagonia
Gulf of San Jorge
Taitao Penninsula
Andes Mountains
Patagonia
Port Stanley
FALKLAND ISLANDS
Tierra Del Fuego
Puerto Toro
SOUTH GEORGIA

Europe—Political and Physical

EUROPE IS THE WORLD'S SECOND-SMALLEST CONTINENT, covering roughly 4 million sq miles (10.4 million km²). It is classed as a continent despite the fact that it is part of the Asia landmass. It contains a wide range of landscape and scenery, with the North European Plain stretching around 2,500 miles (4,023 km) as an area of unbroken lowland. To the south, the Plain is bordered by the mountains of the Alps and the Pyrenees and eroded plateaus or massifs, while to the north there are the older mountains of Scandinavia and northern Great Britain. Western Europe has mainly a temperate climate which changes to a more extreme continental climate both in the east and inland. The north of the continent tends to be colder with tundra and subarctic conditions, while the south has a mediterranean climate with hot, dry summers and warm, but wet winters. Sprawling coniferous forests can be found in the mountainous regions.

Europe is home to both the world's smallest country—The Vatican City—and also (partly) the largest—Russia (which spans from Europe and into Asia). European history has been dominated by shifting and changing political borders. Recent changes have occurred since the breakup of the former Soviet Union and Yugoslavia—both changes seeing a raft of new independent states created. Over 70 percent of the 700 million people living in Europe live in urban areas, and Europe is one of the most densely populated regions of the world. In contrast, there are many small rural settlements found in the more isolated, mountainous parts of the continent.

Europe's large, densely populated lowland areas are highly industrialized and home to many of the world's leading companies. This has also led to a range of environmental incidents over the years, such as the Chernobyl nuclear explosion of 1986 and the recent Hungarian toxic sludge spill. The land is also intensively farmed—with both arable and livestock farming—in order to feed the largely urbanized population.

10 THINGS TO REMEMBER

1. **How many countries are there in Europe?** 50 countries make up the continent of Europe.

2. **How many people live there?** About 731 million people live in Europe.

3. **What languages are spoken there?** There are more than 40 main languages spoken across Europe, originating from the Indo-European language family.

4. **What is the largest city in Europe?** Moscow, Russia, is the largest city in Europe, with some 8 million inhabitants.

5. **What makes Europe unique?** Europe is also home to the world's smallest country, The Vatican City, and also the world's largest, Russia.

6. **What is the largest university in Europe?** La Sapienza University in Rome, Italy, with over 184,000 students.

7. **What is the longest river in Europe?** The longest river in Europe is the River Volga, at 2,300 miles (3,701 km).

8. **What is the highest point?** The highest point is Mount Elbrus in Russia, at 18,510 feet (5,642 m) above sea level.

9. **What is the lowest point?** The lowest point is the Caspian Sea, at 91 feet (28 m) below sea level.

10. **What is the largest lake?** The largest lake is Lake Ladoga in Russia, at 6,800 sq miles (17,611 km²).

COLORING TASK

Color the countries of Europe to match the color plate on page 115, and color the main physical features (such as rivers and mountains).

Africa—Political and Physical

AFRICA IS THE WORLD'S SECOND-LARGEST CONTINENT, covering around 11.7 million sq miles (30.3 million km²) and around 22 percent of Earth's surface. It is surrounded by the Atlantic Ocean to the south and west, and by the Indian Ocean to the east. To the north, it is separated from Europe by the Mediterranean Sea. A large part of northern Africa is dominated by the slowly expanding Sahara Desert, while southern Africa is home to the Great Rift Valley and a range of mountain plateaus.

Fifty-four countries make up the continent of Africa, of which 32 are among the poorest in the world. High levels of debt still hinder development for many African nations. Despite this, some African countries are making significant economic progress, and the richest country in Africa in 2009 was Equatorial Guinea, which had a GDP per capita of $30,200, while South Africa still has by far the biggest economy. Many West African countries rely on cash crops such as coffee and cacao as their main source of income. Libya and Nigeria both have large oil reserves, and many other countries such as Botswana rely on mining of raw materials as their main source of income.

With Africa stretching from the Tropic of Cancer in the north to the Tropic of Capricorn in the south, the continent has a mainly tropical climate with high temperatures all year round. However, there is a wide range of rainfall across the continent, with very little rainfall in the hot, arid desert regions contrasted with the wet regions of central Africa. Environmental issues in Africa are mostly related to the very hot and arid conditions in parts of the continent. Marginal areas on the edges of the deserts are becoming degraded as wood is sought for fuel and the soil becomes exposed, windblown, and eroded.

10 THINGS TO REMEMBER

1. **How many countries are there in Africa?** 54 countries make up the continent of Africa.

2. **How many people live there?** About 1 billion people live in Africa.

3. **What languages are spoken there?** There are around 2,000 different languages used across Africa.

4. **What is the largest city in Africa?** Lagos, Nigeria is the largest city in Europe, with just over 9 million inhabitants.

5. **What makes Africa unique?** Africa is believed to be the birth place of humanity around 5 million years ago.

6. **The first recorded dominant civilization in Africa . . .** was the Egyptians in 3,300 BC.

7. **What is the longest river in Africa?** The longest river in Africa is the River Nile, at 4,160 miles (6,694 km).

8. **What is the highest point?** The highest point is Mount Kilamanjaro in Tanzania, at 19,340 feet (5,895 m) above sea level.

9. **What is the lowest point?** The lowest point is the Lake Assal in central Djibouti, at 511 feet (156 m) below sea level.

10. **What is the largest lake?** The largest lake is Lake Victoria, at 26,600 sq miles (68,893 km²).

COLORING TASK

Color the countries of Africa to match
the color plate on page 115, and color
the main physical features (such as
rivers and mountains).

ASIA IS THE LARGEST OF THE SEVEN CONTINENTS of the world, extending around 5,300 miles (8,529 km) from the Ural Mountains in the west, through Russia and China down to Sumatra and the Philippines in the southeast. It covers over 17.2 million sq miles (44.5 km²) and has a land area larger than the surface of the moon. The mountains and plateaus of the north are much older than the more well-known features of the south. The Himalayas are both the youngest and highest fold mountain range in the world and are the dominant geological feature of the southern part of the continent. The Himalayas began forming about 70 million years ago as the Indian subcontinent crashed into Asia.

There are 49 countries that make up Asia. It is accepted that the continent begins at the Ural Mountains and Turkey in the west, dividing Russia between Europe and Asia. Within Russia, it is almost as if there are two countries, with Moscow and St. Petersburg having heavy European influences. Cities in the Far East, such as Vladivostok, are much more oriented toward China. The north of Russia reaches into the Arctic Circle. To the southwest the continent encompasses the countries of the Middle East, including Saudi Arabia and Israel. Toward the eastern edge of Asia are the islands of Japan, Indonesia, and the Philippines as the continent stretches down toward the equator. Asia has about 60 percent of the world's population, with China and India accounting for a large proportion of these people—both countries have populations in excess of 1 billion people!

Asia experiences a wide range of climates due to its range of latitudes. Much of the north of the continent is subarctic and tundra, while moving south the temperatures increase, but precipitation rates vary greatly. Large parts of the continent are dry, with extensive hot desert regions, while tropical or monsoon climates dominate the coastal regions and the islands to the south.

10 THINGS TO REMEMBER

1. **How many countries are there in Asia?** 49 countries make up the continent of Asia.

2. **How many people live there?** About 4 billion people live in Asia.

3. **What languages are spoken there?** There are a wide variety of different languages used across Asia, and it is estimated that there are over 2,100 different tongues.

4. **What is the most populated city in Asia?** Mumbai, India is the largest city in Asia, with over 12 million inhabitants.

5. **What makes Asia unique?** Asia is home to the giant panda.

6. **What are the three dominant financial centers in Asia?** Tokyo, Hong Kong, and Singapore are the three dominant financial centers.

7. **What is the longest river in Asia?** The longest river in Asia is the Chang Jiang (Yangtze), at 3,915 miles (6,300 km).

8. **What is the highest point?** The highest point is Mount Everest on the border of Nepal and China (Tibet), at 29,035 feet (8,850 m) above sea level.

9. **What is the lowest point?** The lowest point is the Dead Sea shore, at 1,368 feet (417 m) below sea level.

10. **What is the largest lake?** The largest lake is the Caspian Sea, at 143,192 sq miles (370,866 km²).

ARCTIC OCEAN

Chukchi Sea · Bering St.

Zemlya Frantsa Iosifa

Severnaya Zemlya

Novosibirskiye Ostrova

East Siberian Sea

Bering Sea

Barents Sea

Novaya Zemlya

Kara Sea

Laptev Sea

Indigirka

Kolyma

Kolyma Lowland

●Murmansk

●Dikson

Nordvik●

Central Siberian Plateau

Sea of Okhotsk

St. Petersburg★

Pechora

Ob

Siberian Lowland

Yenisey

Lena

Lensk★

Ninhnyaya Tunguska

R U S S I A

Moscow★

Volga

Ural Mountains

Tobol

Ob

Yenisey

Angara

Lena

Khabarovsk●

Volgograd★●

Ural

Ishim

Irtysh

Omsk●

Novosibirsk●

Astana★●

Irkutsk★

Sayan Mts.

Lake Baykal

Manchurian Plain

Harbin●

Sapporo●

Black Sea

Istanbul●

Caucasus Mts.

Caspian Sea

KAZAKHSTAN

Altai

Ulaanbaatar★

M O N G O L I A

Changchun●

Sea of Japan

JAPAN

Ankara★

TURKEY

Anatolian Plateau

ARMENIA

T'bilisi★

AZERBAIJAN

Aral Sea

Lake Balkhash

Gobi Desert

NORTH KOREA

Tokyo★

Nicosia★

Yerevan★

UZBEKISTAN

Syr Darya

Bishkek★

Tien Shan

Beijing★

Pyongyang★

Seoul★

Osaka●

LEBANON

SYRIA Beirut★

Baku★

TURKMENISTAN

Tashkent★

KYRGYZSTAN

Tarim Basin

SOUTH KOREA

ISRAEL Damascus★

Amman★

JORDAN

Tigris

Baghdad★

Euphrates

Tehran★

Ashkhabad★

Dushanbe★

TAJIKISTAN

Yellow Sea

Sea of Japan

PACIFIC

OCEAN

Jerusalem★

IRAQ

I R A N

Kunlan Shan

Xi'an●

Great Basin

Wuhan●

Shanghai●

East China Sea

SAUDI Kuwait★

KUWAIT

Zagros Mts.

AFGHANISTAN

Hindu Kush

Kabul★

Plateau Of Tibet

C H I N A

ARABIA Kuwait City

Al Manamah★

BAHRAIN

Islamabad★

Himalayas

Mt. Everest▲

Chang Jiang (Yangtze)

Ad Dawhah★ QATAR

PAKISTAN

NEPAL

Kathmandu★

Thimpu★

BHUTAN

Taipei●

Red Sea

Riyadh★

Abu Dhabi★

U.A.E.

Indus

New Delhi★

Ganges

Muscat★

Karachi●

Deccan

Calcutta●

Dhaka★

Hong Kong●

Sanaa★

YEMEN

OMAN

Arabian Sea

I N D I A

MYANMAR

Irrawaddy

Naypyidaw★

LAOS

Hanoi★

South China Sea

Gulf of Aden

Mumbai●

Hyderabad●

BANGLADESH

Bay of Bengal

Vientiane★

Hue●

Da Nang●

Manila★

Salween

THAILAND

VIETNAM

PHILIPPINES

Philippine Sea

Moulmein●

Chennai●

Bangkok★

CAMBODIA

Laccadive Islands (to India)

Phnom Penh★

Ho Chi Minh●

Mekong

Celebes Sea

Colombo★

SRI LANKA

Bandar Seri Begawan★

BRUNEI

MALDIVES

Malé★

Kuala Lumpur★

M A L A Y S I A

I N D O N E S I A

Coral Sea

Singapore City★

SINGAPORE★

Pontianak●

Jakarta★

Dili★

EAST TIMOR

I N D I A N

OCEAN

E A S T   I N D I E S

COLORING TASK

Color the countries of Asia to match the color plate on page 115, and color the main physical features (such as rivers and mountains).

Australia and Oceania—Political and Physical

THE CONTINENT OF AUSTRALIA AND OCEANIA (also sometimes referred to as Australasia and Oceania) includes Australia, New Zealand, and the groups of islands in the southern and central Pacific Ocean. Despite the 3.3 million sq miles (8.5 million km²) of land throughout the continent, Australia and Oceania only contains 0.5 percent of the world's population (about 35 million people). There are more than 10,000 Pacific islands offering a great diversity of environments, from almost barren, waterless, coral atolls to vast, continental islands. Australia itself is a tectonically stable island, but much of the rest of the continent, including New Zealand, lies on, or close to, plate boundaries with many islands volcanic in origin.

Australia is the largest of the 23 countries that make up Australia and Oceania. It includes few other countries with significant land areas apart from New Zealand, Papua New Guinea, and Fiji. The wide range of Pacific islands is split into the three main groups of Micronesia, Melanesia, and Polynesia. Australia is the world's sixth-largest country and is flat, dry, and sparsely populated. Over two-thirds of the country is hot and arid land, classed as outback, and most of the population resides in the more temperate areas near the coast. Much of the continent was settled by European colonial powers such as Great Britain and France, but most countries have now achieved independence.

With the continent spreading across a large part of the Pacific Ocean, Australia and Oceania has a wide range of climate types. Many of the Pacific islands enjoy a tropical climate, while the center of Australia is hot desert and New Zealand experiences a temperate climate, similar to that of much of northern Europe.

10 THINGS TO REMEMBER

1. **How many countries are there in Australia and Oceania?** 23 countries make up the continent.

2. **How many people live there?** About 35 million people live in Australia and Oceania.

3. **What languages are spoken there?** There is a wide range of traditional languages spoken, although English, French, and Spanish have become common since colonization.

4. **What is the largest city in Australia and Oceania?** Sydney, Australia is the largest city in Australia and Oceania, with just over 4 million inhabitants.

5. **What makes Australia and Oceania unique?** It has some unique animal life including the koala bear, platypus, and kangaroo.

6. **What is the longest river in Australia and Oceania?** The River Murray, at 1,476 miles (2,375 km).

7. **What is the highest point?** The highest point is Mount Wilhelm in Papua New Guinea, at 14,793 feet (4,509 m) above sea level.

8. **What is the lowest point?** The lowest point is the Lake Eyre shore, at 52 feet (16 m) below sea level.

9. **What is the largest lake?** The largest lake is the Lake Eyre, at 3,668 sq miles (9,500 km²).

10. **What is the longest reef?** The Great Barrier Reef stretches along the eastern coast of Australia for more than 1,429 miles (2,300 km).

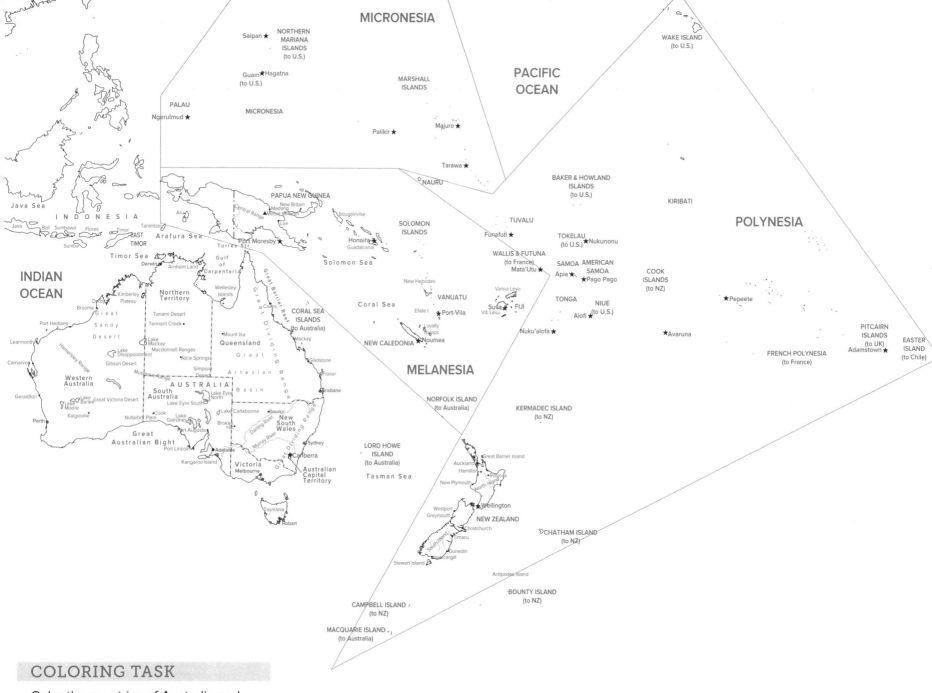

MICRONESIA

Saipan ★ NORTHERN
MARIANA
ISLANDS
(to U.S.)

Guam ★ Hagatna
(to U.S.)

PALAU

Ngerulmud ★

MICRONESIA

PACIFIC
OCEAN

WAKE ISLAND
(to U.S.)

MARSHALL
ISLANDS

Palikir ★

Majuro ★

Tarawa ★

○ NAURU

BAKER & HOWLAND
ISLANDS
(to U.S.)

KIRIBATI

POLYNESIA

Java Sea

INDONESIA

Java Bali Sumbawa Flores
Sumba Timor
EAST
TIMOR

Arafura Sea
Torres Str.

PAPUA NEW GUINEA

Central Range
Madang New Britain
Mount William
Port Moresby
Lae

Aru

Bougainville

SOLOMON
ISLANDS

Honiara
Guadalcanal

Solomon Sea

TUVALU

Funafuti ★

TOKELAU
(to U.S.)
★ Nukunonu

WALLIS & FUTUNA
(to France)
Mata'Utu ★

SAMOA AMERICAN
Apia ★ SAMOA
★ Pago Pago

COOK
ISLANDS
(to NZ)

★ Pepeete

INDIAN
OCEAN

Timor Sea

Darwin

Arnhem Land

Northern
Territory

Kimberley
Plateau

Broome

Great

Sandy

Desert

Tanami Desert

Tennant Creek

Gulf
of
Carpentaria

Wellesley
Islands

Cairns

CORAL SEA
ISLANDS
(to Australia)

Coral Sea

New Hebrides

VANUATU

Efate I. ★ Port-Vila

Loyalty
Islands

NEW CALEDONIA

Noumea

Vanua Levu

Suva FIJI
Viti Levu

TONGA

Nuku'alofa ★

NIUE
(to U.S.)
Alofi ★

★ Avaruna

FRENCH POLYNESIA
(to France)

PITCAIRN
ISLANDS
(to UK)
Adamstown ★

EASTER
ISLAND
(to Chile)

MELANESIA

Learmonth

Port Hedland

Carnarvon

Geraldton

Perth

Hamersley Range

Western
Australia

Lake
Barlee
Lake
Moore

Kalgoorlie

Nullarbor Plain

Broome

Derby

Lake
Disappointment

Gibson Desert

Musgrave Range

AUSTRALIA
South
Australia

Great Victoria Desert

Cook
Port Augusta

Port Lincoln

Kangaroo Island

Mount Isa

Queensland

Lake
Mackay

Macdonnell Ranges

Alice Springs

Simpson
Desert

Great

Artesian

Basin

Lake Eyre
North

Lake Eyre South

Lake
Gairdner

Lake Callabonna

Bourke

Broken
Hill

New
South
Wales

Darling River

Murray River

Adelaide

Victoria

Melbourne

Mackay

Gladstone

Fraser

Brisbane

Great Dividing Range

Great Barrier Reef

Sydney

Australian
Capital
Territory

Canberra

NORFOLK ISLAND
(to Australia)

LORD HOWE
ISLAND
(to Australia)

KERMADEC ISLAND
(to NZ)

Tasman Sea

Auckland
Hamilton
New Plymouth

Great Barrier Island

Rotorua

North Island

Westport
Greymouth

Wellington

NEW ZEALAND

Christchurch

South Island

Timaru

Dunedin
Invercargill

Stewart Island

CHATHAM ISLAND
(to NZ)

Tasmania

Hobart

CAMPBELL ISLAND
(to NZ)

Antipodes Island

BOUNTY ISLAND
(to NZ)

MACQUARIE ISLAND
(to Australia)

Great
Australian Bight

COLORING TASK

Color the countries of Australia and
Oceania to match the color plate on
page 116, and color the main physical
features (such as rivers and mountains).

Antarctica—Political and Physical

ANTARCTICA IS EARTH'S SOUTHERNMOST CONTINENT and is the location of the South Pole. It is situated almost entirely south of the Antarctic Circle and is surrounded by the Southern Ocean. At 5.4 million sq miles (14 million km²), it is the fifth-largest continent in area after Asia, Africa, North America, and South America. About 2 percent of Antarctica is not covered by permanent ice, which averages at least 5,000 feet (1,524 m) in thickness.

Antarctica, on average, is the coldest, driest, and windiest continent and has the highest average elevation of all the continents. Antarctica is considered a desert, with annual precipitation of only 8 inches (20.3 cm) along the coast and far less inland. There are no permanent human residents, but anywhere from 1,000 to 5,000 people, mainly scientists, reside throughout the year at the research stations scattered across the continent. Only cold-adapted plants and animals survive there, including penguins, seals, whales, fish, and tundra vegetation on the Antarctic Peninsula.

Although legends exist about a mythical great southern continent and date back as early as the first century AD, the first confirmed sighting of the continent was in 1820 by a Russian expedition. The coasts of the continent, however, remained largely unexplored for the rest of the nineteenth century largely because of its hostile environment and remote location. It wasn't until 1911 that the Norwegian Roald Amundsen became the first man to reach the South Pole.

The first formal use of the name "Antarctica" as a continental name in the 1890s is attributed to the Scottish cartographer John George Bartholomew. The name Antarctica is derived from a Greek word meaning "opposite to the north."

10 THINGS TO REMEMBER

1. Antarctica is the fifth-largest continent and makes up 10 percent of Earth's land area.

2. Only 2 percent of the land in Antarctica is not covered in ice.

3. Antarctic ice, which at its thickest reaches 3 miles (5 km) in depth, comprises almost 70 percent of Earth's fresh water. If it all melted, sea levels would rise by about 164 feet (50 m).

4. Due to its ice cap, Antarctica is the highest continent, averaging 7,545 feet (2,300 m) above sea level, and the highest peak is Vinson Massif, at 16,076 feet (4,900 m).

5. Antarctica has the lowest recorded temperature; -130°F (-90°C) at Vostock in 1983. Inland, temperatures range from -94°F (-70°C) in the winter to -31°F (-35°C) in the summer.

6. Antarctica is the windiest place on Earth, with gusts up to 203 mph (327 km/h) having been recorded.

7. Antarctica is the driest place on Earth. In some places like the Dry Valleys, it has not rained for thousands of years.

8. Around 270 million years ago, Antarctica was part of Gondwanaland and probably covered with tundra, marsh, and forests, explaining why coal and petrified wood can still be found today.

9. Antarctica is the least known of Earth's landmasses; fewer than 200,000 people have ever been there.

10. Scientists from all over the world go to Antarctica to study such things as the organisms that live in the unspoiled ecosystem and the consequences of climate change.

■ Year-round research station

Twenty-one of 28 Antarctic consultative nations have made no claims to Antarctic territory (although Russia and the United States have reserved the right to do so), and they do not recognize the claims of the other nations.

DID YOU KNOW?

The Antarctic Treaty was drafted and signed by 12 countries in 1959, prohibiting military activity and mineral mining so that the continent's ecosystem is protected. As of 2009, 46 countries have signed the agreement.

COLORING TASK

Color the map and color including the main physical features (such as rivers and mountains).

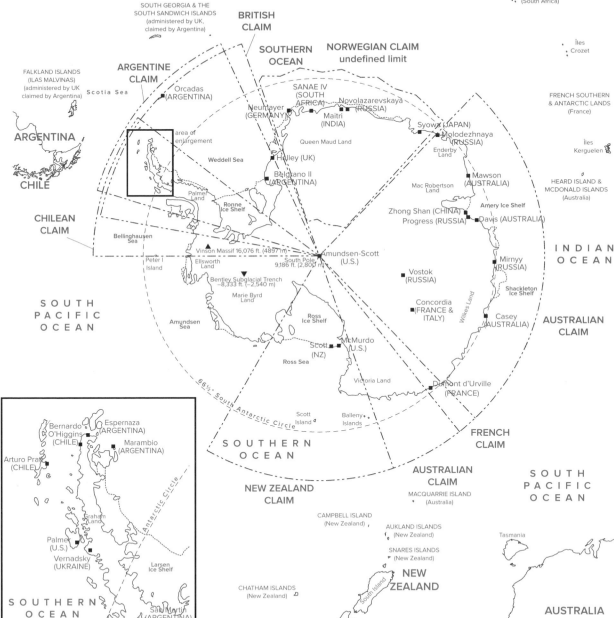

SOUTH AFRICA

SOUTH ATLANTIC OCEAN

Antarctic Convergence

BOUVET ISLAND
(Norway)

PRINCE EDWARD ISLANDS
(South Africa)

Îles Crozet

SOUTH GEORGIA & THE SOUTH SANDWICH ISLANDS
(administered by UK, claimed by Argentina)

BRITISH CLAIM

SOUTHERN OCEAN

NORWEGIAN CLAIM
undefined limit

FRENCH SOUTHERN & ANTARCTIC LANDS
(France)

FALKLAND ISLANDS
(ILAS MALVINAS)
(administered by UK claimed by Argentina)

ARGENTINE CLAIM

Scotia Sea

Orcadas (ARGENTINA)

SANAE IV (SOUTH AFRICA)

Novolazarevskaya (RUSSIA)

Neumayer (GERMANY)

Maitri (INDIA)

Syowa (JAPAN)

Molodezhnaya (RUSSIA)

Îles Kerguelen

ARGENTINA

CHILE

area of enlargement

Queen Maud Land

Enderby Land

Mawson (AUSTRALIA)

HEARD ISLAND & MCDONALD ISLANDS
(Australia)

Weddell Sea

Halley (UK)

Belgrano II (ARGENTINA)

Mac Robertson Land

CHILEAN CLAIM

Palmer Land

Ronne Ice Shelf

Zhong Shan (CHINA)
Progress (RUSSIA)

Amery Ice Shelf

Davis (AUSTRALIA)

INDIAN OCEAN

Bellinghausen Sea

Peter I Island

Ellsworth Land

Vinson Massif 16,076 ft. (4897 m)

South Pole 9,186 ft. (2,800 m)

Amundsen-Scott (U.S.)

Mirnyy (RUSSIA)

Shackleton Ice Shelf

Bentley Subglacial Trench −8,333 ft. (−2,540 m)

Marie Byrd Land

Vostok (RUSSIA)

SOUTH PACIFIC OCEAN

Amundsen Sea

Ross Ice Shelf

Concordia (FRANCE & ITALY)

Wilkes Land

Casey (AUSTRALIA)

AUSTRALIAN CLAIM

McMurdo (U.S.)

Scott (NZ)

Ross Sea

Victoria Land

Dumont d'Urville (FRANCE)

FRENCH CLAIM

66½° South Antarctic Circle

Scott Island

Balleny Islands

SOUTHERN OCEAN

AUSTRALIAN CLAIM

MACQUARRIE ISLAND
(Australia)

SOUTH PACIFIC OCEAN

NEW ZEALAND CLAIM

CAMPBELL ISLAND
(New Zealand)

AUKLAND ISLANDS
(New Zealand)

SNARES ISLANDS
(New Zealand)

Tasmania

CHATHAM ISLANDS
(New Zealand)

NEW ZEALAND

AUSTRALIA

North Island

South Island

Bernardo O'Higgins (CHILE)

Espernaza (ARGENTINA)

Marambio (ARGENTINA)

Arturo Prat (CHILE)

Graham Land

Palmer (U.S.)

Vernadsky (UKRAINE)

Larsen Ice Shelf

San Martin (ARGENTINA)

SOUTHERN OCEAN

Rothera (UK)

Antarctic Circle

World Time Zones

SINCE EARTH IS LIKE A GIANT BALL, it is difficult to locate places on the globe. A system of lines has been developed in order to pinpoint any place on the surface of Earth. Lines of latitude run in an east-west direction and allow us to say how far we are north or south. Lines of latitude are equally spaced horizontal lines shown on globes and maps. These lines are often referred to as parallels because they are all aligned in the same direction. There are 89 of these lines to the north of the equator (the line of latitude around the middle of Earth) and 89 to the south. Where the ninetieth line would be are the two points of the North and South poles.

Lines of longitude run from north to south and allow us to say where we are east or west. They are the vertical lines shown on globes and maps. These lines run around Earth from the North Pole to the South Pole, dividing the world into 360°. The prime meridian is the line of longitude that runs through Greenwich, UK, at 0°. There are 180° of longitude to the west of the prime meridian and 180° to the east. The opposite of the prime meridian is often referred to as the Antimeridian. The Antimeridian is important because time changes by an entire day when you cross this point—therefore, the International Date Line was created. The International Date Line does not follow the 180° Antimeridian exactly, as it swerves around the nations and islands of the Pacific Ocean.

Sir Sanford Fleming, a Scottish-born Canadian, proposed the notion of world time zones in the late nineteenth century. He suggested that world time zones were set to try and ensure that each country in the world experiences noon at the time when the sun is at its highest point in the sky. He noted that Earth rotates at the rate of 15° every hour, and it would therefore be beneficial to divide the world into 24 equal time zones of 15° each and to adjust clocks accordingly in each time zone. His system was finally adopted in 1929 and is arguably the most important system of the modern world.

10 THINGS TO REMEMBER

1. Some places/countries use time offsets that are not an integral number of hours from GMT such as Newfoundland (Canada), which is GMT -03:30 (summer -02:30) and Afghanistan, which is +04:30.

2. Usual Daylight Saving Time (summertime) rule is to alter the clocks ahead by one hour.

3. Australia has both horizontal and vertical time zones in the summer.

4. Prior to 1995, the International Date Line split the country of Kiribati. The result was that the eastern part of Kiribati was a whole day and two hours behind the western part of the country where its capital is located. In 1995, the International Date Line was moved east, placing the entire country into the same day.

5. Equatorial and tropical countries usually do not observe Daylight Saving Time as the duration of day and night are very nearly always the same, at 12 hours.

6. Although Russia is geographically spread over 12 time zones, it officially observes only 9 time zones.

7. China observes one uncommonly wide time zone GMT +08:00.

8. Brazil sets its summertime by decree every year. Some states and counties observe summertime on a year-to-year basis.

9. Some countries use different rules to start and end DST. For example, a law in Israel requires that summer must last at least 150 days.

10. Greenwich meantime has the same time as London time during the winter; however, London is one hour ahead of GMT during summertime.

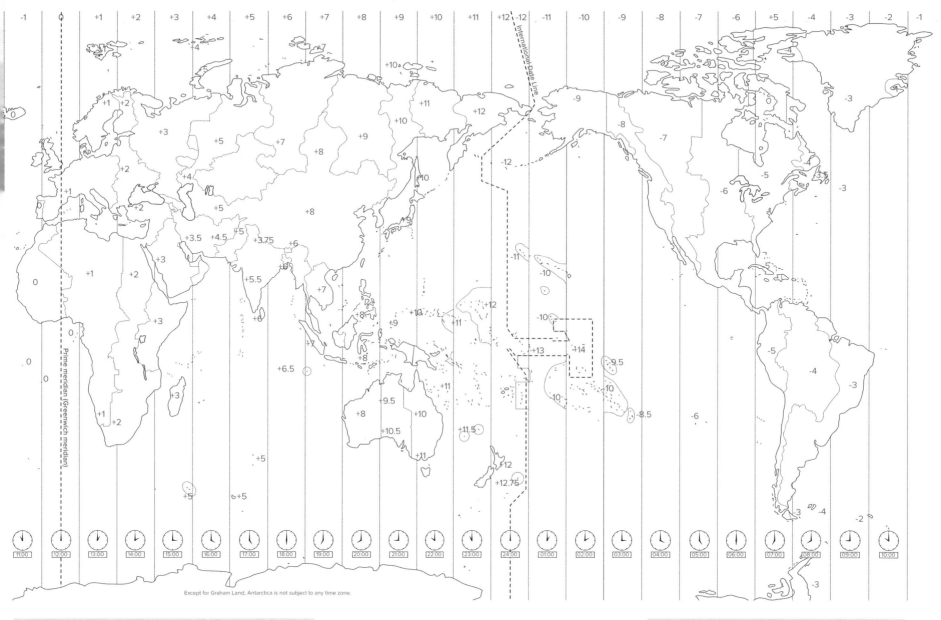

Except for Graham Land, Antarctica is not subject to any time zone.

MEASURING TIME

The Greenwich Meridian
The Greenwich meridian was chosen as the prime meridian of the world in 1884 at a meeting of 41 delegates from 25 nations in Washington, D.C., at the International Meridian Conference. Greenwich won the prize of longitude 0° by a vote of 22 to 1 against (San Domingo), with two abstentions (France and Brazil).

COLORING TASK

Color each of the 24 time zones. These can be broken down into two groups of 12 moving west and east from the prime meridian. Each of the time zones will need to be numbered so that they can be accurately colored.

The United States

THE UNITED STATES IS THE WORLD'S THIRD-LARGEST COUNTRY in size and nearly the third-largest in terms of population. It borders Canada to the north and Mexico to the south. It is made up of 50 states, of which two are separated from the mainland area—Alaska, northwest of Canada, and the islands of Hawaii in the Pacific Ocean. Today, the U.S. remains the world's only real superpower, with a strong industrial economy and its people generally enjoying a high quality of life.

From 13 original British colonies, the United States has grown into a country that covers around 3.5 million sq miles (9 million km²), with a wide range of natural environments and populations. All 50 states share sovereignty with the federal government, but have their own constitution, legislature, judiciary, executive branch, and capital city.

The United States was the first country to seek independence from the British Empire. The 13 colonies began a rebellion against British rule in 1775, and this culminated with the Declaration of Independence on July 4, 1776, although the war with Great Britain dragged on until 1783. George Washington was elected the first president of the United States in 1789. The expansion of the United States began in 1791, when Vermont was admitted, while the Mexican War during the 1840s added Florida, Texas, Iowa, Wisconsin, and California, and this continued until Alaska and Hawaii were admitted as the forty-ninth and fiftieth states in 1959.

DID YOU KNOW?
THE 13 ORIGINAL COLONIES

The original 13 colonies were British colonies on the East Coast of North America. They were Connecticut, Delaware, Georgia, Maryland, Massachusetts, New Hampshire, New Jersey, New York, North Carolina, South Carolina, Pennsylvania, Rhode Island, and Virginia. By 1774, these colonies had well-established systems of self-government and elections, although they sought protection from what they called "Imperial Interference" from the British. These Colonies formed the birth place of the United States. when they were granted independence in 1776.

10 THINGS TO REMEMBER

1. **California** was the first state to ever reach a trillion-dollar economy in gross state product.

2. **Colorado** is the only state in history to turn down the Olympics. In 1976, the Winter Olympic Games was planned to be held in Denver. Sixty-two percent of all state voters chose at almost the last minute not to host the Olympics, because of the cost, pollution, and population boom it would have on the state of Colorado and the city of Denver.

3. **Hawaii** is the most isolated population center on the face of Earth. It is 2,390 miles (3,846 km) from California; 3,850 miles (6,195 km) from Japan; 4,905 miles (7,895 km) from China; and 5,280 miles (8,497 km) from the Philippines.

4. Cheeseburgers were first served in 1934 at Kaolin's restaurant in Louisville, **Kentucky**.

5. The **Mississippi** River is the largest in the United States and is the nation's main waterway. Its nickname is "Old Man River."

6. **Nebraska** was once called "The Great American Desert."

7. **New York state** is home to 58 species of wild orchids.

8. In 1946, Philadelphia, **Pennsylvania**, became home to the first computer.

9. **Texas** is the only state to have the flags of six different nations fly over it. They are: Spain, France, Mexico, Republic of Texas, Confederate States, and the United States.

10. The state of **Washington** is the only state to be named after a United States president.

COLORING TASK

Color each of the 10 states named
in the 10 Things to Remember a
separate color.

Physical Features of the United States

THE UNITED STATES OF AMERICA BORDERS BOTH THE NORTH ATLANTIC AND NORTH PACIFIC OCEANS and the countries of Canada and Mexico. It is the third-largest country in the world by area and has a varied topography. The eastern regions consist of hills and low mountains, while the central interior is a vast plain (called the Great Plains region), and the west has high rugged mountain ranges (some of which are volcanic in the Pacific Northwest). Alaska also features rugged mountains as well as river valleys. The U.S. can be divided into seven main physiographic divisions, which are, running from east to west: the Atlantic–Gulf Coastal Plain; the Appalachian Highlands; the Interior Plains; the Interior Highlands; the Rocky Mountain System; the Intermontane Region; and the Pacific Mountain System. An eighth division, the Laurentian Uplands, a part of the Canadian Shield, dips into the United States from Canada in the Great Lakes region.

The Atlantic coast of the U.S. is relatively low lying with the Atlantic plain. The exception is the Appalachian Mountain range, which cuts through, running from just north of Atlanta to the southern edge of Pittsburgh. The western Pacific coast is by contrast much more hilly and mountainous with a much greater relief than the Atlantic coast.

The climate of the U.S. varies depending on location. It is considered mostly temperate, but is tropical in Hawaii and Florida, arctic in Alaska, semiarid in the plains west of the Mississippi River, and arid in the Great Basin of the Southwest. Temperatures can range from 132°F (56°C) at the height of summer in Death Valley, California, to -79°F (-62°C) in the extreme winter in Alaska. Northern states of the U.S. are the coldest, often experiencing bitter, freezing winters—especially in the plains of the Midwest and Northeast. However, the "Sunbelt" southern states remain warm all year and temperatures rarely drop below freezing.

10 THINGS TO REMEMBER

1. The U.S. covers nearly 3.8 million sq miles (9.8 million km²).

2. Alaska has the longest coastline in the U.S., at 6,640 miles (10,686 km).

3. The lowest temperature ever recorded in the U.S. was -79°F (-62°C) at Prospect Creek in Alaska in January 1971.

4. The highest temperature ever recorded in the U.S. was 132°F (56°C) in Death Valley, California in July 1913.

5. The deepest lake in the U.S. is Crater Lake, Oregon. At 1,932 feet (589 m), it is the world's seventh-deepest lake.

6. The tallest mountain in the world is located in Mauna Kea, Hawaii, and is only 13,795 feet (4,205 m) above sea level, however, when measured from the sea floor it is over 32,808 feet (10,000 m) high, making it taller than Mount Everest.

7. There are about 4,000 miles (6,437 km) of navigable inland channels in the U.S.

8. The mean elevation of the U.S. is 2,503 feet (763 m) above sea level.

9. The geographic center of the 50 states is 44°58'N lat.103°46'W longitude in Butte County, South Dakota.

10. There are 20 designated world heritage sites in the U.S., including the Everglades National Park in Florida and the Grand Canyon National Park in Arizona.

COLORING KEY

1. >8,202 feet
   (2,500 m)

2. 4,921–8,201 feet
   (1,500–2,499 m)

3. 1,640–4,920 feet
   (500–1,499 m)

4. 656–1,639 feet
   (200–499 m)

5. 0–655 feet
   (0–199 m)

COLORING TASK

Color the height of land areas of the U.S. using the five parameters in the coloring key to your right.

Political Map of the United States

THE U.S. IS THE WORLD'S LARGEST and most technologically advanced economy, made up mainly of industrial and service sectors. The main industries include motor vehicles, telecommunications, petroleum, steel, aerospace, chemicals, electronics, food processing, consumer goods, lumber, and mining. Although agricultural production is only a small part of the economy, a variety of foodstuffs are produced such as corn, wheat and other grains, vegetables, fruit, cotton, beef, pork, poultry, dairy products, fish, and forest products. Oil is a crucial commodity for the U.S.'s economy and the state of Alaska has large resources of oil. This oil is transported via a 1,000-mile (1,600-km) pipeline from Northern Alaska to the port of Valdez along the Trans-Alaskan pipeline.

The map shows both the main cities of the U.S. as well as the key automobile transport routes. The U.S. is heavily dependent on transportation by road and has nearly 4 million miles (6.4 million km) of paved roadway with the highest number of motor vehicle ownership in the world (estimated to be at 772 motor vehicles per 1,000 people).

The U.S. government is made up of a representative democracy with two legislative bodies. These bodies are the Senate and House of Representatives. The Senate is made up of 100 seats with two members from each of the 50 states, while the House of Representatives consists of 435 seats which are elected by the people from the 50 states. An elected president of the U.S. is both Head of Government and Chief of State.

DID YOU KNOW?
WASHINGTON, D.C.

Washington, District of Columbia, is the capital of the United States and was founded on July 16, 1790. It is distinct from the 50 states because it serves as a separate federal district and the permanent capital. The city of Washington was originally a separate municipality within the federal territory until a single unified municipal government for the whole district was established in 1871. The city is named in honor of the first American president, George Washington.

10 THINGS TO REMEMBER

1. The population of the U.S. is 310,232,863 (July 2010 estimate).

2. Population density in the U.S. is 31.8 people per 0.3 sq mile (1 km²).

3. About 80 percent of the people in the U.S. live in urban areas, leaving just 20 percent in the large rural areas.

4. The capital city is Washington, D.C.

5. The largest city in the U.S. is New York City with just over 8 million inhabitants.

6. Six of the 25 tallest buildings in the world are in the U.S.

7. Life expectancy in the U.S. is:

   White males—73.8 years
   White females—79.6 years
   Non-white males—68.9 years
   Non-white females—76.1 years

Source: 2008 data taken from www.america.gov

8. 15.5 percent of the U.S. population isn't covered by private or government health insurance.

9. There are over 48 million visitors to the United States each year.

10. There are 100 million single Americans, making up 44 percent of all U.S. residents over the age of 15.

COLORING TASK

Color the map to match
the color plate at the back
of the book.

World Climate Zones

THE WORLD HAS SEVERAL CLIMATIC ZONES, which are summarized on the map to the right. The atmospheric systems of the world are driven by the energy of the sun, and the distribution of this energy and its variation cause the different climate zones that have emerged across the planet. The tropical climates found in a belt around the equator are separated from the two polar regions by a large temperate zone. The factors that influence the climate zones are:

Distance from the equator (latitude): As you move away from the equator, temperature decreases, while the temperature range increases. The curved surface of Earth means that the sun's rays are dispersed over a larger area of land as you move away from the equator. The polar regions are much colder because the sun's rays have to travel much farther to reach them.

Height above sea level (altitude): Temperature decreases with height, and the less dense air found at altitude cannot hold heat as easily as at sea level.

Winds: Winds blown from hot areas (such as deserts) will carry warm air and will raise temperatures, while if they have been blown from cold areas (such as the poles), they will lower temperatures.

Distance from the sea: The sea can have a cooling effect in the summer and a warming effect in the winter. This is because land heats and cools at a faster rate than sea water. Therefore, coastal areas tend to have mild winters and cool summers, while inland areas tend to have higher temperatures in the summer and colder temperatures in the winter.

Aspect: Slopes that face the sun will be warmer than those that do not. Therefore, in the Northern Hemisphere, southern facing slopes tend to be warmer, while in the Southern Hemisphere, northern slopes tend to be the warmest.

10 THINGS TO REMEMBER

1. **Highest recorded temperature:** 136°F (58°C) at Al Aziziyah, Libya, on September 13, 1922.

2. **Lowest recorded temperature on a permanently inhabited continent:** -90°F (-68°C) at Verkhoyansk, Siberia, on February 6, 1933.

3. **Driest place:** Quillagua, Chile, which had 0.01 inches (0.5 mm) of rainfall between 1964 and 2001.

4. **Wettest place (over 12 months):** Cherrapunji, northeast India, between August 1860 and August 1861. Cherrapunji also holds the record for the most amount of rainfall in one month, with 115 inches (2,930 mm) of rainfall in July 1861.

5. **Highest recorded wind speed:** 230 mph (371 km/h) on April 12, 1934 at Mt. Washingston in New Hampshire, U.S.

6. **Windiest place:** With regular gales of over 198 mph (320 km/h), Commonwealth Bay in Antarctica is considered to be the windiest place on Earth.

7. **Highest barometric pressure:** 1,083 mb on December 31, 1968 at Agata, Siberia.

8. **Lowest barometric pressure:** 870 mb on October 12, 1979 during Typhoon Tip, which occurred 298 miles (480 km) west of Guam in the Pacific Ocean.

9. **Sunniest place:** Yuma, Arizona, gets more than 4,000 hours of sunshine per year, making it the sunniest place on the planet!

10. **Least sunny place:** The South Pole is the least sunny place, receiving only 182 days of sunshine a year.

COLORING TASK

Color each of the six main climate zones onto the world map using the number key.

COLORING KEY

1. ☐ Polar—Very cold and dry year-round

2. ☐ Temperate—Warm summers and cold winters

3. ☐ Arid—Very dry and hot year-round

4. ☐ Tropical—Remains hot and wet throughout the year

5. ☐ Mediterranean—Dry, hot summers with mild, cool winters

6. ☐ Mountains (tundra)—Remains very cold year-round

Tectonic Plates

EARTH'S CRUST IS NOT A CONTINUOUS, UNBROKEN SHELL. It is divided into several major, and a number of smaller, tectonic plates that move. The movement of these tectonic plates is called "continental drift," and there is evidence that the tectonic plates have moved over long periods of time:

- Some continents fit almost perfectly together, such as South America and Africa;
- Similar fossils have been found on different continents, showing that they were once joined together;
- Almost identical patterns of rock layers found on different continents are further evidence that they were once joined together.

Earth's surface is made up of a number of large plates that are in constant slow motion. This is because the tectonic plates making up the crust of Earth are "floating" on a liquid mantle (made of hot liquid rock). Convection currents in the mantle move the plates. It is at the weaker edges of these plates, the plate boundaries where the plates meet, where earthquakes and volcanoes occur.

There are four types of plate boundaries that can be found where the plates meet:

Destructive Plate Margin: This is where one plate slides beneath another as they collide. The bottom plate crumples, creating new mountains and volcanoes.

Constructive Plate Margin: This is where two plates are moving apart from each other. Molten rock from the mantle rises to the surface, cools, and hardens, forming a ridge of new rock.

Conservative Plate Margin: This is where two plates are sliding past each other. Pressure builds up until they move with a "jerk," causing earthquakes.

Collision Plate Margin: This is where two plates collide and are crushed against each other. They are pushed upward and form new mountains.

10 THINGS TO REMEMBER

EARTHQUAKES

1. Earthquakes occur when vast plates, or rocks, within Earth suddenly break or shift under stress, sending shock waves rippling.

2. Earthquake waves are measured on sensitive instruments called seismographs.

3. The Richter scale assigns earthquakes a number based on the power of their seismic waves.

4. Thousands of earthquakes occur every day around the globe, although most of them are too weak to be felt.

5. Every year about 10,000 people, on average, die as a result of earthquakes.

VOLCANOES

1. Volcanoes are vents in Earth's surface from which molten rock, debris, and steam erupt.

2. About 1,900 volcanoes are active today or are known to have been active in historical times.

3. An eruption begins when magma, the molten rock from deep within Earth's crust, rises toward the surface.

4. Although some volcanoes are considered extinct, almost any volcano is capable of rumbling to life again.

5. Volcanoes provide valuable mineral deposits, fertile soils, and geothermal energy.

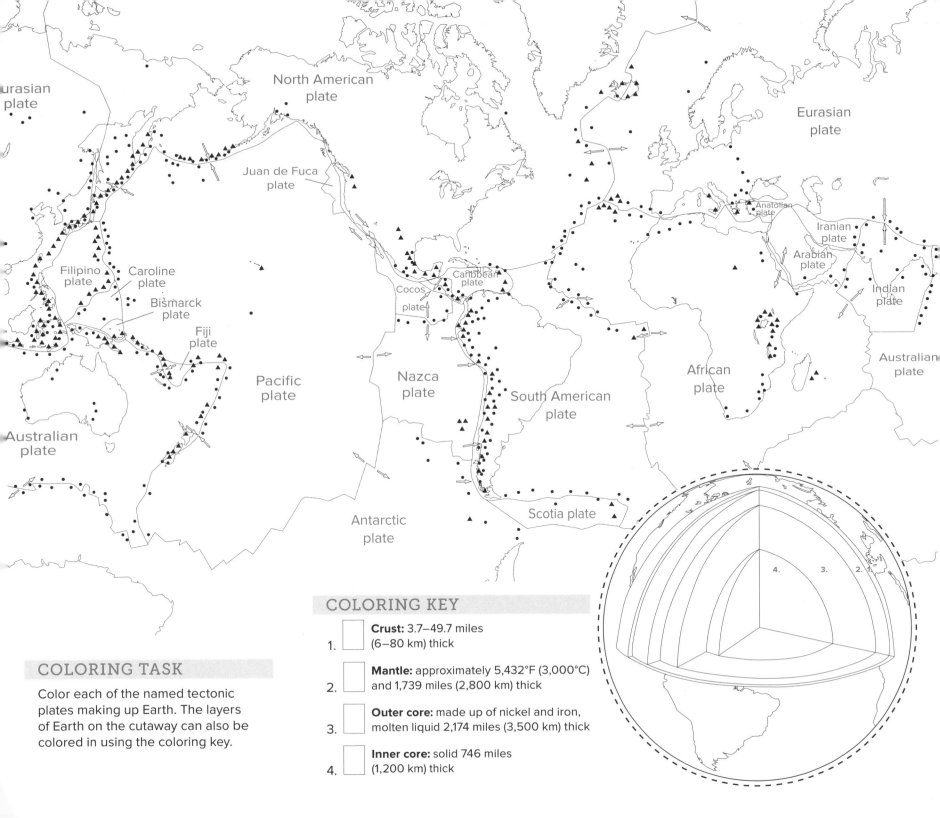

North American plate

Eurasian plate

Eurasian plate

Juan de Fuca plate

Anatolian plate

Iranian plate

Arabian plate

Filipino plate

Caroline plate

Bismarck plate

Fiji plate

Indian plate

Cocos plate

Caribbean plate

Pacific plate

African plate

Australian plate

Australian plate

Nazca plate

South American plate

Antarctic plate

Scotia plate

COLORING KEY

1. Crust: 3.7–49.7 miles (6–80 km) thick

2. Mantle: approximately 5,432°F (3,000°C) and 1,739 miles (2,800 km) thick

3. Outer core: made up of nickel and iron, molten liquid 2,174 miles (3,500 km) thick

4. Inner core: solid 746 miles (1,200 km) thick

COLORING TASK

Color each of the named tectonic plates making up Earth. The layers of Earth on the cutaway can also be colored in using the coloring key.

Features of a Volcano

VOLCANOES ARE CAUSED BY THE MOVEMENT OF EARTH'S TECTONIC PLATES. They can occur at both constructive and destructive plate margins. A volcano is a mountain that is formed as an opening at a plate boundary and acts as a vent where molten magma forces its way from deep inside Earth to the surface. Magma can appear as flows of molten lava, as volcanic bombs, as fragments of rock, or even simply, as ash and dust. Volcanoes can be considered to be active, dormant, or extinct:

- A volcano is considered to be active if it has erupted recently and is likely to erupt again in the future. There are thought to be about 700 active volcanoes around the world such as Mount Etna in Italy.
- Dormant (or sleeping) volcanoes are those that may have erupted in the past 2,000 years, but have not done so in recent history. These volcanoes can be dangerous as it can be difficult to predict when, or if, they will erupt again. Mount St. Helens in Washington state, in the U.S., was thought to be dormant until it erupted again in 1980.
- A volcano is classed as extinct if its volcanic activity has ceased and it is thought unlikely to ever erupt again. Mount Snowdon in Wales, UK, was once an active volcano some 50 million years ago and is now extinct.

The map on the right shows the locations of some of the world's major volcanoes and where volcanic activity can occur. Volcanoes occur in long narrow belts around the world matching the plate margins. These belts include:

- The Pacific "Ring of Fire" which encircles the Pacific Ocean stretching all along the West Coast of both South and North America, up past Alaska, across to the tip of Russia, and down through Southeast Asia to New Zealand.

10 THINGS TO REMEMBER

1. Volcanic ash is very good for soil, so plants can grow well in the period after a volcanic eruption.

2. Volcanic slopes left after an eruption can be very steep, so rare and delicate plants and animals can set up home there and be protected.

3. One in 10 people in the world live within the "danger range" of an active volcano.

4. There are around 1,510 "active" volcanoes in the world.

5. The biggest volcano in the world is Mauna Loa in Hawaii. Its whole volume is about 19,193 cubic miles (80,000 km^3).

6. Sometimes lightning is seen in volcanic clouds.

7. Volcanoes can change the weather.

8. Volcanoes can also have long-term effects on the climate, making the world cooler.

9. Fast-moving lava can kill people, and falling ash can make it hard for them to breathe. Famine, fires, and earthquakes resulting from volcanoes can also add to the mortality rates.

10. Lava can kill plants and animals, too. The Mount St. Helens volcano in 1980 killed an estimated 24,000 animals, including 11,000 hares, 6,000 deer, 300 bobcats, 200 black bears, and 15 cougars.

Heimaey

Vesuvius

Etna

Mt. St. Helens

Fuji

Mauna Loa

Popocatapeti

Paricutin

Nevada del Ruiz

Cotopaxi

akatoa

Ngauruhoe

Aconcagua

Steam, gas, and dust

Volcanic bombs

Falling ash

Crater

Lava flow

Secondary cone

Layers of
ash and lava

Main vent

Magma chamber

COLORING TASK

Color each land mass that has
volcanoes present.

Features of an Earthquake

EARTHQUAKES ARE CAUSED BY SMALL MOVEMENTS of Earth's tectonic plates. They occur deep under the surface of Earth, but the tremors can often be felt on the surface of Earth. The point where the earthquake begins under the Earth is called the focus, while the point on the surface of Earth directly above the focus is called the epicenter. Earthquakes can occur at any of the tectonic plate margins. An earthquake happens when two of Earth's plates get locked up with the pressure of movement between them both, causing enormous tension to build up. Eventually some of the rock gives way causing the plates to move. The tension between the plates is released, and waves of energy (called seismic waves) travel in all directions. The earthquake is the vibrations that these waves cause. As the plates settle into their new positions, there can be many smaller earthquakes which are referred to as "aftershocks."

The map on the following page shows there is a clearly identifiable pattern of where earthquakes are distributed in long, narrow belts in parts of the world. These belts include:

- The belt that stretches across southern Europe and Asia, linking the Pacific and Atlantic oceans
- The belt that runs down the entire length of the Atlantic Ocean following the line of the mid-Atlantic trench
- The belt that encircles the Pacific Ocean following the "Ring of Fire"

Earthquakes are measured using a machine called a "seismometer" which can detect very small movements of Earth. The magnitude of an earthquake is measured on the Richter scale. An increase of 1 on this scale means a 10-fold increase in the amplitude of the waves and a 30-fold increase in their energy. To date, no earthquake has measured more than 8.9 on the Richter scale.

10 THINGS TO REMEMBER

1. The property damage price tag for the quake was over $5.9 billion.
2. An estimated 3,757 people were injured, and there were 67 reported fatalities.
3. The Goodyear blimp, present for the World Series game, ended up providing aerial shots of the damage.
4. This earthquake is sometimes referred to as the "World Series Quake" because it interrupted the third game of the series which was postponed for 10 days as a result.
5. The actual quake was broadcast on a live television feed.
6. This earthquake lasted 15 seconds at a magnitude of 6.9.
7. Although most of the damage occurred in San Francisco, Oakland, and Santa Cruz, damage was reported in many other locations in the region, including Hayward, San Mateo, Monterey, and Boulder Creek.
8. Over 18,300 homes and 2,575 businesses were reported damaged.
9. The earthquake left an estimated 12,000 people homeless.
10. The Forest of Nisene Marks State Park in Santa Cruz was the epicenter of the earthquake.

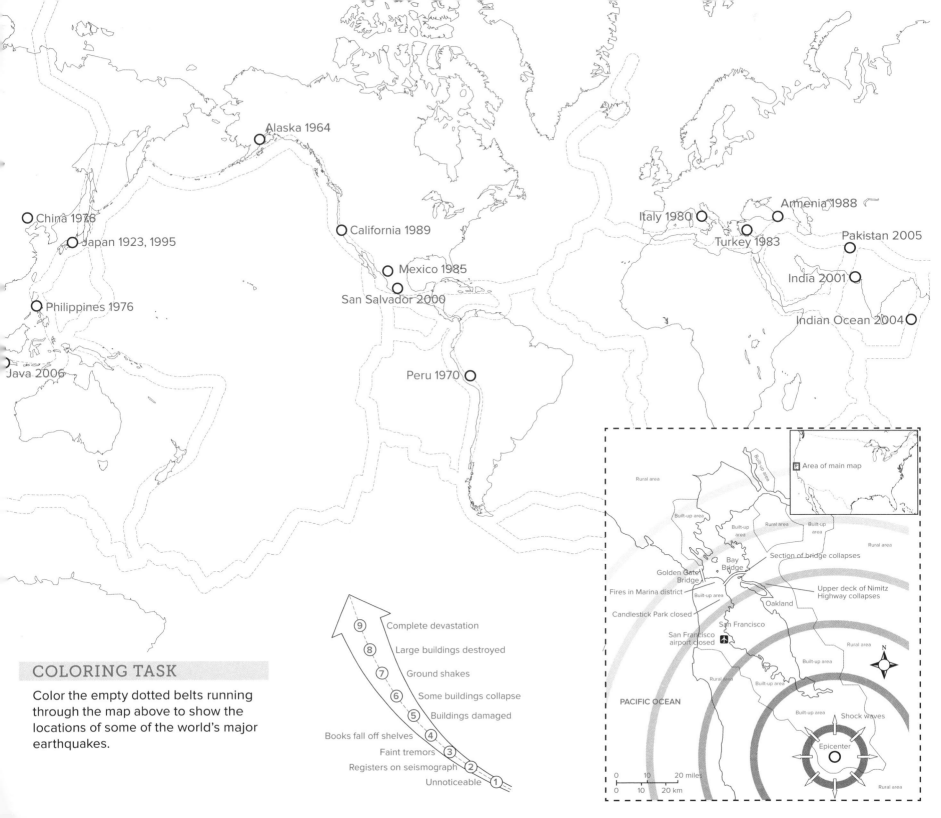

Alaska 1964

China 1976

Japan 1923, 1995

California 1989

Mexico 1985

San Salvador 2000

Philippines 1976

Italy 1980

Armenia 1988

Turkey 1983

Pakistan 2005

India 2001

Indian Ocean 2004

Java 2006

Peru 1970

COLORING TASK

Color the empty dotted belts running through the map above to show the locations of some of the world's major earthquakes.

9 Complete devastation
8 Large buildings destroyed
7 Ground shakes
6 Some buildings collapse
5 Buildings damaged
4 Books fall off shelves
3 Faint tremors
2 Registers on seismograph
1 Unnoticeable

Rural area

Built-up area

Area of main map

Built-up area

Built-up area

Rural area

Rural area

Built-up area

Rural area

Section of bridge collapses

Bay Bridge

Golden Gate Bridge

Upper deck of Nimitz Highway collapses

Fires in Marina district

Built-up area

Oakland

Candlestick Park closed

San Francisco airport closed

San Francisco

Rural area

Built-up area

Rural area

Built-up area

PACIFIC OCEAN

Built-up area

Shock waves

Epicenter

0   10   20 miles
0   10   20 km

Rural area

The Hydrological Cycle

WATER IS WHAT MAKES EARTH UNIQUE from other planets in the solar system. Without it, life could not exist—it is therefore more important than anything else on Earth! The hydrosphere consists of all the water on Earth—in seas, oceans, rivers, lakes, rocks, and soil, in living things, and in the atmosphere. Water exists on Earth's surface in three states: as a liquid (water), as a solid (ice), and as a gas (water vapor). Water can change states by:

- Evaporation—liquid changing to vapor
- Condensation—vapor turning into liquid
- Melting or freezing

Of all the world's water, 97 percent is contained in the seas and oceans as salt water and is therefore unsuitable for use by humans, animals, and terrestrial plants. That leaves just 3 percent of fresh water, of which 2 percent is locked up as ice and snow in arctic and alpine areas, leaving just 1 percent of all the world's water as fresh water on land, or water vapor in the atmosphere. Fresh water can be found naturally occurring on Earth's surface in bogs, ponds, lakes, rivers, and streams, and underground as groundwater in aquifers and underground streams.

The hydrological cycle (or water cycle) is the continuous and never-ending movement of water between the land, the sea, and the atmosphere. The hydrological cycle is a closed system—the water continues to go around and around, but none of it is ever lost from the system as a whole, so Earth never gets wetter or drier.

There are a number of stores in the global hydrological cycle, such as in rocks, soil, or lakes, and oceans, and the water can remain in these stores for varying amounts of time—from around just one day, to thousands of years! In the hydrological cycle these stores are linked by processes which transfer water into and out of them, and these processes regulate the cycle.

10 KEY WORDS RELATING TO THE HYDROLOGICAL CYCLE

1. **Precipitation:** All the water released from clouds such as rain, snow, hail, or sleet.
2. **Infiltration:** The movement of water into the soil from the surface.
3. **Groundwater discharge:** The movement of water below the water table.
4. **Groundwater storage:** Water which is stored underground.
5. **Evaporation:** When water which is heated by the sun becomes vapor and rises into the atmosphere. This can take place over land or sea.
6. **Surface runoff:** Water flowing across the surface. The water may be a channel, such as a river or a stream.
7. **Sublimation:** The evaporation of ice to water vapor, without passing through a liquid state.
8. **Evapotranspiration:** The process of evaporation from the water that plants lose through their leaves.
9. **Condensation:** When water vapor is cooled and turns into water droplets to form clouds.
10. **Transpiration:** The process whereby water contained in liquid form in plants is converted to vapor and released to the atmosphere.

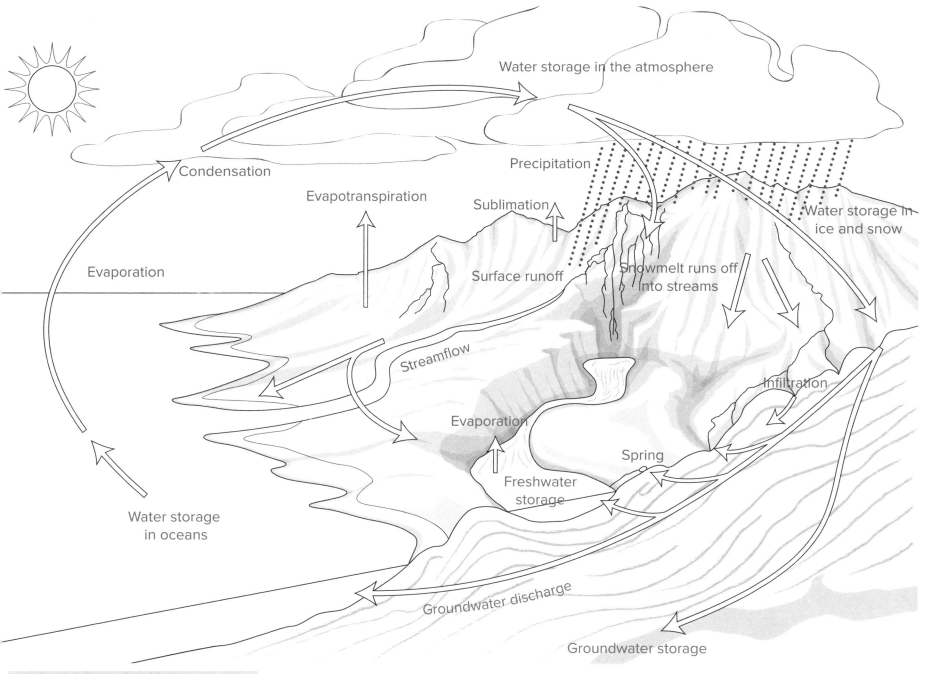

Water storage in the atmosphere

Condensation

Evapotranspiration

Sublimation

Precipitation

Water storage in ice and snow

Evaporation

Surface runoff

Snowmelt runs off into streams

Streamflow

Infiltration

Evaporation

Spring

Water storage in oceans

Freshwater storage

Groundwater discharge

Groundwater storage

COLORING TASK

Color in the path of water in the hydrological cycle illustrated above. Begin with "Precipitation" at the top of the diagram.

The River Basin System

A RIVER BASIN IS PART OF THE HYDROLOGICAL CYCLE, but unlike the hydrological cycle, it is an open system. This means that it has both inputs into the system through precipitation (rain and snow) and outputs from the system where water is lost from the system either by rivers carrying it to the sea, or through evapotranspiration. Within the river basin system there are stores and transfers:

- **Stores** are places where the water is held such as lakes and ponds on the surface, or soil and rocks underground
- **Transfers** are processes by which the water moves through the system, such as surface runoff, or infiltration

A drainage basin is the area of land drained by a river and its tributaries. Large rivers such as the Amazon or the Mississippi have vast drainage basins covering many thousands of miles. The boundary of a drainage basin is called the watershed, and this is usually a range of hills or mountains. Rain that falls beyond the watershed will fall into another river and is part of another drainage basin.

When it rains, most water droplets will be intercepted by trees and plants. If the rain falls as a short shower, then little water will reach the ground. It will be stored on leaves and then lost to the system through evapotranspiration. However, if the rainfall is heavier, it will drip from the leaves to the ground where it may form pools (surface storage). As the ground becomes wetter, some of the water will infiltrate through the soil. The water will then either be stored in the soil or slowly transferred sideways (throughflow) or downward (percolation). Percolation forms groundwater which is stored deep below the surface in rocks and can lead to groundwater flow. Surface runoff can occur when the soil is saturated and no more water can infiltrate the soil.

10 THINGS TO REMEMBER

1. The Nile River in Africa is the longest in the world.
2. The Rhine and Rhone rivers both begin in Switzerland.
3. The Amazon River in South America has the most outflow of water.
4. The Murray/Darling River is the longest in Australia.
5. The Colorado River in the U.S./Mexico no longer reaches the sea.
6. The Volga River is the largest in Europe.
7. The Rio Grande forms much of the border between the U.S. and Mexico.
8. The Han River flows through Seoul, South Korea.
9. The St. Lawrence River empties the Great Lakes.
10. The Swan River flows through Perth, Australia.

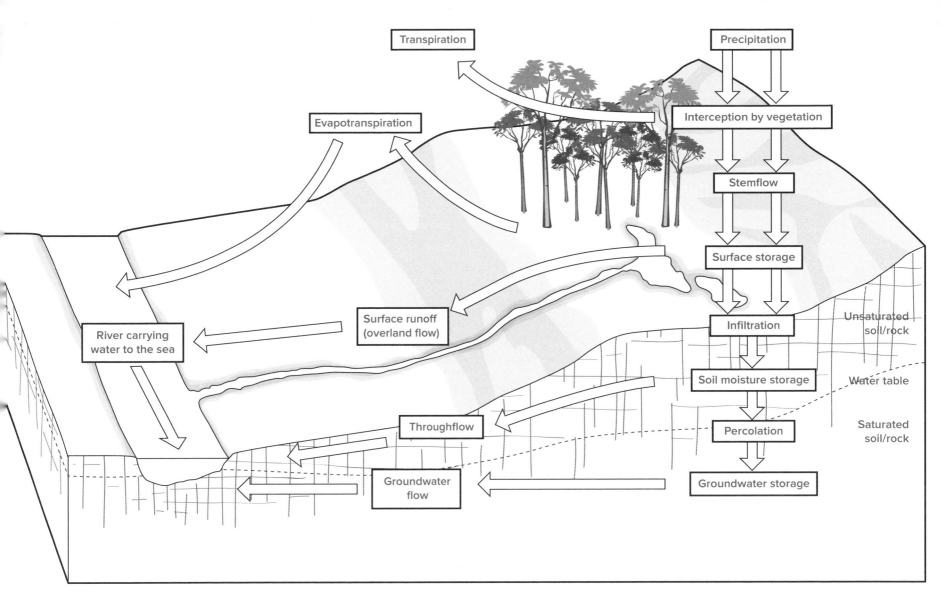

Transpiration

Precipitation

Evapotranspiration

Interception by vegetation

Stemflow

Surface storage

Surface runoff
(overland flow)

River carrying
water to the sea

Infiltration

Unsaturated
soil/rock

Soil moisture storage

Water table

Throughflow

Percolation

Saturated
soil/rock

Groundwater
flow

Groundwater storage

COLORING TASK

Color in the path of the river basin
system illustrated above. Begin
with "Precipitation" at the top of the
diagram.

River Erosion and Deposition

RIVERS TRANSPORT WATER FROM HIGH GROUND TO LOW GROUND. Along the course of a river there are many changes that can occur. A river can be broken down into three main sections:

- **The upper course:** Rivers begin their journey in hills or mountains. The river will flow quickly, carrying large amounts of sediment, and it will flow around interlocking spurs of higher land. There is a lot of bedload and many large rocks are angular in shape. Waterfalls are a common feature in this part of the river.

- **The middle course:** At this stage of a river, the land will now be much flatter, many streams will have joined the river making it much wider. During wet weather, the river may burst its banks and spread onto the floodplain on either side of the river. Meanders are a common feature at this stage of a river.

- **The lower course:** As the river approaches the end of its journey, it moves much faster, carrying a large quantity of water from all the streams and rivers that have fed into it. The river tends to widen, forming an estuary. This part of the river can be affected by the tide and is a mixture of fresh water and salt water. Deposition of fine silt and clay around the estuary can form extensive mud flats and salty marshes.

10 KEY WORDS RELATING TO RIVER EROSION AND DEPOSITION

1. **Drainage Basin:** The area from which rain water drains into the river.

2. **Watershed:** An imaginary line that separates one drainage basin from the next.

3. **Confluence:** The point where two rivers join.

4. **Floodplain:** The land that gets flooded when a river overflows.

5. **Mouth:** This is where the river flows into a lake, sea, or ocean.

6. **Source:** This is the starting point of the river. It could be a spring, a melting glacier, or just an area where a lot of rain collects.

7. **Tributaries:** These are smaller rivers or streams that flow into the main river.

8. **Channel:** This is what the river flows in.

9. **Riverbank:** This is the side of the river channel.

10. **Riverbed:** This is the bottom of the river where material can collect.

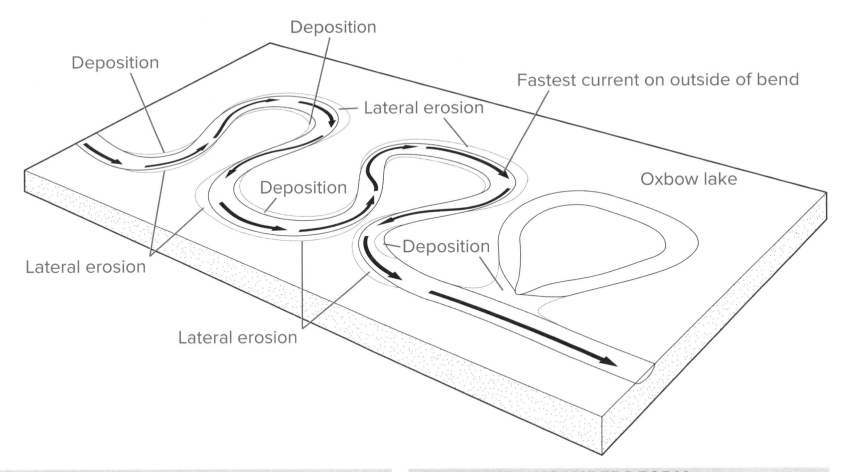

Deposition

Deposition

Deposition

Lateral erosion

Fastest current on outside of bend

Oxbow lake

Lateral erosion

Deposition

Lateral erosion

CROSS SECTION OF A RIVER

Small river cliff

Fastest current

Slowest current

Slipoff slope

Sand and shingle deposited

Outside bank undercut by lateral erosion

HOW RIVER MEANDERS FORM

Fast water on the outside bend of a river causes erosion (or wearing away of the riverbank), while slower water on the inside bend causes deposition (the dropping of material) to occur. The continued deposition on the inside bend leads to a slipoff slope forming that extends into the river, pushing the water on the outside bend to erode the riverbank further. Sometimes these bends in the river can erode into each other forming a straight channel and leaving a separate lake of water known as an "oxbow" lake.

COLORING TASK

Color in the course of the river, and mark the areas of erosion and deposition.

Glaciers

GLACIERS ARE LIKE GREAT RIVERS OF ICE. As they slowly descend, they sculpt mountains and carve out valleys. Many glaciers around the world continue to flow and shape the land in many places today.

A glacier is formed high up in mountainous areas when snow falls. This is called the "zone of accumulation." Air is trapped between the flakes of snow as they fall, this air is compressed and the air is squeezed out, causing the snow to become firmer. As more snow accumulates, the underlying layers are compressed to form a whitish grain like snow called "firn." The weight of more snow falling forces out more air, and the granules of firn merge together to form bluish glacier ice (ice with no air in it turns blue). The glacier will move slowly down the side of a mountain under the force of gravity.

As a glacier moves down a mountainside, it will erode material that it comes in contact with. There are two main processes of glacial erosion:

- **Plucking:** This is where the ice in the glacier freezes onto pieces of rock as it moves over them and picks the rock up and carries it with the glacier as it moves down the mountain.
- **Abrasion:** This is when the pieces of rock that have been picked up by the glacier scrape along at the bottom of the glacier as it moves, eroding the side of the mountain, often leaving carved out marks that are visible on the mountain once the glacier has retreated.

It is possible to see evidence of the existence of glaciers in the past due to the material that is left behind when a glacier has retreated, or melted. There are five types of "moraines," that is the material that a glacier deposits: Terminal Moraine, Lateral Moraine, Medial Moraine, Recessional Moraine, and Ground Moraine.

TOP-10 GLACIERS TO VISIT IN THE WORLD

1. **Perito Moreno Glacier, Argentina:** Located 48 miles (77 km) from the Argentine town of El Calafate, it is one of only three of the southern Patagonian glaciers that are still advancing.

2. **Glacier Bay, Alaska:** Home to more than 100,000 glaciers!

3. **Furtwängler Glacier, Mount Kilimanjaro, Tanzania:** Of the world's receding glaciers, Mount Kilimanjaro's are among the most highly publicized. During the last century, they have lost more than 80 percent of their mass.

4. **Pasterze Glacier, Austria:** The largest of Austria's 925 glaciers.

5. **Vatnajökull Glacier, Iceland:** Due to Iceland's volcanic activity, there are hot springs located within the glacier's ice caves.

6. **Yulong Glacier, China:** It has been receding steadily since the early 1980s.

7. **Fox and Franz Josef, New Zealand:** Although they retreated for much of the twentieth century, they are currently advancing. They are unique among the world's glaciers in that they extend down the mountains and into a temperate rainforest.

8. **Athabasca Glacier, Canada:** It has already lost half its volume and continues to recede.

9. **Biafo Glacier, Pakistan:** Located in the Karakoram Mountain Range.

10. **Antarctica:** Tourists are heading to Antarctica's countless mountains, bays, and glaciers in record numbers—nearly 35,000 now visit each year.

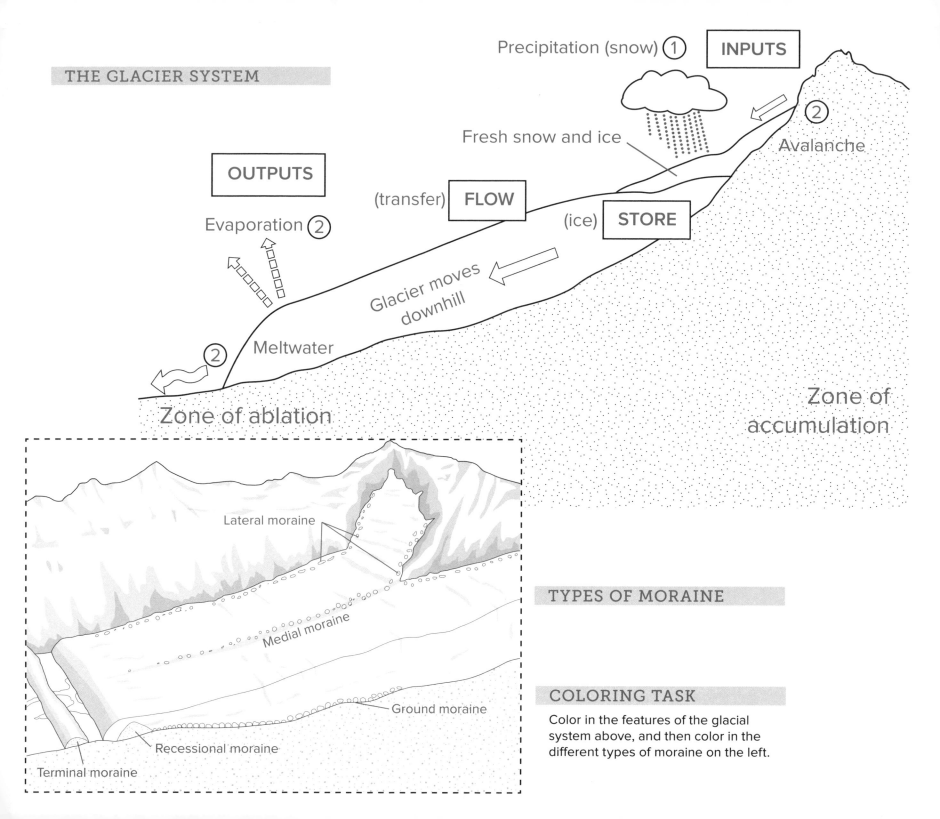

THE GLACIER SYSTEM

Precipitation (snow) ① INPUTS

Fresh snow and ice

Avalanche ②

OUTPUTS

Evaporation ②

(transfer) FLOW

(ice) STORE

Glacier moves downhill

Meltwater ②

Zone of ablation

Zone of accumulation

TYPES OF MORAINE

Lateral moraine

Medial moraine

Ground moraine

Recessional moraine

Terminal moraine

COLORING TASK

Color in the features of the glacial
system above, and then color in the
different types of moraine on the left.

Coastal Processes

A COAST IS A NARROW AREA OF CONTACT ZONE between the land and the sea. The coast is shaped and changed by the effects of the land, air, and sea processes. Some coasts are low lying with long sandy beaches, while others can be steep with rocky cliffs. They have many human uses, including ports, fishing, industry, or tourism. The topography (or shape) of a coast depends upon the geology (or rock type) or relief (height of the land) where it is found. More resistant rocks such as granite and chalk usually form coasts with high, steep cliffs. Less resistant rocks, like clays, erode more easily. Low, flat land often has wide, open beaches.

Waves: Waves from the sea can impact on the coastline. Waves are formed by the wind blowing over the sea and large waves form when there are strong winds. The distance of water over which the wind can blow is called the "fetch." There are two types of waves:

- **Constructive waves** will form if the shore is shallow, and the wave spills forward for a long distance. These waves push material up onto a beach
- **Destructive waves** will form if the shore is steep causing the wave to plunge down and hit the shore with great force. These waves erode material from the beach

Longshore Drift: Material that is eroded from the cliffs by the sea is worn down by the process of attrition and is moved by the sea to be deposited farther along the coast. This process is called "longshore drift" in which material is carried up a beach by waves at the angle that the waves are blown in by (this is rarely at right angles). As the wave pulls back straight down the beach (at a right angle), the material is dragged straight down under gravity and is slowly moved along the coast in a zigzag fashion. Groins can be put in place along beaches to mitigate the effects of longshore drift.

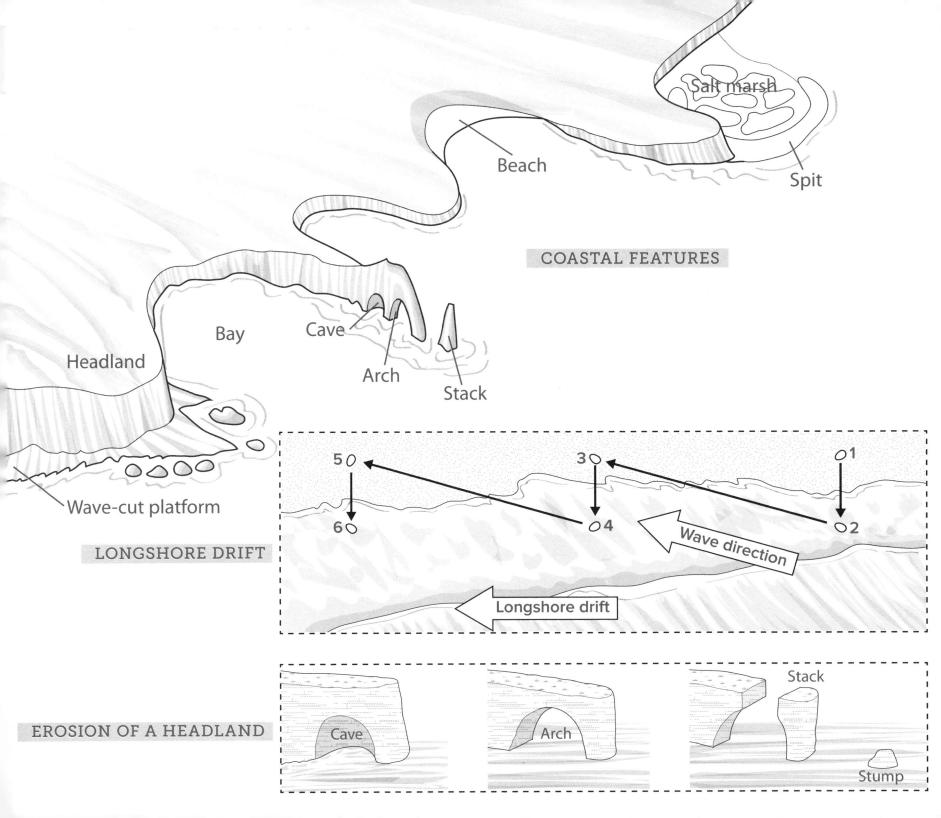

Salt marsh

Beach

Spit

COASTAL FEATURES

Bay

Cave

Headland

Arch

Stack

Wave-cut platform

LONGSHORE DRIFT

5

3

1

6

4

2

Wave direction

Longshore drift

EROSION OF A HEADLAND

Cave

Arch

Stack

Stump

Population Density

IN 2008, THERE WERE JUST OVER 6.7 BILLION PEOPLE living on planet Earth. The six billionth person was born in 1999, and the seven billionth will probably arrive sometime in 2012. During a person's lifetime there will be another one billion people added to the world who will require food, water, and shelter. By 8000 BC, it had taken the human race one million years to double in number. Yet by 1968, the world's population had reached 3.5 billion, and the doubling time had fallen to just 35 years!

The United Nations Population Division expected the world population to grow from 6 billion to 9.3 billion by 2050. It is now expected to reach only 8.9 billion by 2050. This is because people are having fewer babies than expected, especially in developing countries.

The fears of an overpopulated planet seem to be disappearing. In the 1960s, nearly every country on the planet had an expanding population, but from 1971 the number of babies born in developed countries dropped. Many of these countries (such as Sweden and Italy) are now seeing their populations begin to fall slightly. Other countries (such as China and India) introduced controls to limit their populations, and the number of babies being born in developing countries is beginning to level off. With less babies being born overall, the United Nations expects the 8.9 billion figure to be a peak that may slowly begin to drop after the year 2200.

The world's population is not evenly distributed and is affected by physical factors such as relief, climate, vegetation, soils, natural resources, and water supply. However, at a regional scale, the pattern is more influenced by human factors, which may be economic, political, or social.

10 MOST DENSELY POPULATED COUNTRIES IN THE WORLD

Population per sq mile (km²)

1. **Monaco:** 6,229 people per sq mile (16,135.0 km²)
2. **Singapore:** 2,950 people per sq mile (7,641.0 km²)
3. **Malta:** 484 people per sq mile (1,256.0 km²)
4. **Maldives:** 436 people per sq mile (1,131.1 km²)
5. **Bahrain:** 422 people per sq mile (1,095.1 km²)
6. **Bangladesh:** 396 people per sq mile (1,027.9 km²)
7. **Taiwan:** 273 people per sq mile (709.0 km²)
8. **Mauritius:** 254 people per sq mile (660.1 km²)
9. **Barbados:** 250 people per sq mile (649.1 km²)
10. **Nauru:** 235 people per sq mile (610.0 km²)

WORLD POPULATION GROWTH FROM 1500

Year	Number of people
1500	500 million
1804	1 billion
1927	2 billion
1960	3 billion
1974	4 billion
1987	5 billion
1999	6 billion

PROJECTED POPULATION GROWTH

Year	Number of people
2013	7 billion
2028	8 billion
2054	9 billion
2183	10 billion

Source: United Nations, *The World at Six Billion*

1. ☐ > 500 people per 0.39 sq mile

2. ☐ 200–499 people per 0.39 sq mile

3. ☐ 100–199 people per 0.39 sq mile

4. ☐ 50–99 people per 0.39 sq mile

5. ☐ 10–49 people per 0.39 sq mile

6. ☐ < 10 people per 0.39 sq mile

COLORING TASK

Complete a choropleth map of the world using the number key to highlight the population density of each country around the world.

Population Change

ALTHOUGH THE POPULATION OF THE WORLD is expected to continue to grow until about 2200, this growth is not evenly distrubuted around the world. The map on the right shows that:

- **Population increase** is mainly happening in Africa, the Middle East, and parts of South America and South Asia
- **Population balance** is mainly found in North America and Europe
- **Population decline** is happening in Russia and parts of central and eastern Europe

Higher levels of population increase are occurring in developing (or, low-income and middle-income) countries. Lower levels of population increase, population balance, and even population decline mainly occur in developed (or, high-income) countries.

The population of a country is constantly changing. In some countries the population will be growing, while in others it may stay level or may even decline. The change of population in a country is dependent on the number of babies being born and the number of people that will die. The difference between the births and deaths is called the "natural increase." If birth and death rates are almost equal, then it means that the country will have a population balance.

A comparison in 2003 of the future population growth of Yemen with Russia, highlights the changes that can occur in a country. In 1950, Russia's total population was 103 million people, yet in Yemen there were just 4.3 million. In 2000, Russia's population had grown to 145 million and Yemen's population to 17.5 million. However, by 2050, Russia's population is expected to fall to about 104 million, while Yemen's population is expected to rise to 102 million. In just over 40 years the population of Yemen could meet that of Russia and then overtake it!

10 SELECTED COUNTRIES AND THEIR RATE OF NATURAL INCREASE

% natural increase of population (1990–2005)

1. **Brazil:** 1.5
2. **Cambodia:** 2.5
3. **Chile:** 1.4
4. **Democratic Republic of the Congo:** 2.8
5. **Greece:** 0.6
6. **Norway:** 0.6
7. **Portugal:** 0.4
8. **Senegal:** 2.5
9. **Sierra Leonne:** 2
10. **United Kingdom:** 0.3

Source: The World Bank

COLORING TASK

Complete a choropleth map of the world using the number key to highlight the population increase of each country around the world.

COLORING KEY

1. ☐ <0%

2. ☐ 0.0–0.9%

3. ☐ 1.0–1.9%

4. ☐ 2.0–2.9%

5. ☐ >3.0%

Migration to the United States

THE UNITED STATES HAS HAD SOME FORM OF IMMIGRATION POLICY since its founding in 1783. Since 2006, the U.S. has allowed more legal immigrants to settle permanently than any other country in the world. There are now some 37 million foreign-born residents making up 12.4 percent of the total U.S. population. However, at least one third of these are thought to have entered the country illegally. The immigration policy of the U.S. has four main aims:

- To reunite families by admitting family immigrants who already have family members living in the U.S.
- To admit workers with specific skills for jobs where workers are in short supply
- To provide refuge for people who face racial, political, or religious persecution in their country of origin
- To provide greater ethnic diversity by admitting people from countries with previously low rates of immigration to the U.S.

In 1990, the U.S. government passed the Immigration Act. The aim of this was to help businesses attract skilled foreign workers. An annual limit of 675,000 permanent immigrant visas (permits to enter the U.S.) was set. Of these, 55,000 are put into the "green card" visa lottery. It was also agreed that 125,000 refugees would be admitted each year. There are two types of U.S. visas:

- An immigrant visa for people who intend to live and work permanently in the U.S.
- A nonimmigrant visa is for people who live in other countries and wish to stay temporarily in the United States (e.g. tourists, students, or diplomats)

TOP 10 IMMIGRANT COUNTRIES TO THE UNITED STATES

(2006)

1. Mexico (173,753)
2. People's Republic of China (87,345)
3. Philippines (74,607)
4. India (61,369)
5. Cuba (45,614)
6. Colombia (43,151)
7. Dominican Republic (38,069)
8. El Salvador (31,783)
9. Vietnam (30,695)
10. Jamaica (24,976)

COLORING TASK

Complete a choropleth map of the world using the number key to indicate the immigration rate of each sending country to the U.S.

COLORING KEY

Immigration rate relative to total population of sending country, 2001–2005 (rates per thousand) based on 2000 U.S. census.

1. ☐ 0–0.29

2. ☐ 0.30–0.99

3. ☐ 1.00–2.99

4. ☐ 3.00–9.99

5. ☐ 10.00–53.89

Wealth—GDP per Capita

THE GROSS DOMESTIC PRODUCT (GDP) PER CAPITA is the total value of all the goods and services produced by a country in one year divided by the number of people living in that country. It is usually measured in U.S. dollars to allow comparisons between different countries around the world. The measure of GDP per capita allows the economic development of countries to be made and compared.

The map on the right shows that more than 46 percent of the world's wealth is in North America and western Europe. The regions with the highest GDP per capita are North America, Japan, and western Europe. The region with the lowest GDP per capita is central Africa. Taking all the central African states together and adding their GDP, it would account for less than 1 percent of the GDP of the world's richest region, North America. By the same token, if North America redistributed just 1 percent of its wealth, then the GDP of central Africa would more than double.

Wealth, as reflected by GDP per person, is highest in Luxembourg, Norway, and Switzerland. It is lowest in Ethiopia, Burundi, and the Democratic Republic of the Congo.

TOP 10 COUNTRIES WITH HIGHEST GDP PER CAPITA

(in U.S. dollars, 2002)

1.	Luxembourg	47,354
2.	Norway	41,974
3.	Switzerland	36,687
4.	United States	36,006
5.	Denmark	32,179
6.	Japan	31,407
7.	Ireland	30,982
8.	Iceland	29,749
9.	Qatar	28,634
10.	Greenland	27,648

COLORING TASK

Complete a choropleth map of
the world using the number key to
highlight the GDP per capita of each
country around the world.

COLORING KEY

1. ☐ >U.S. $10,000

2. ☐ U.S. $2,500–9,999

3. ☐ U.S. $1,200–2,499

4. ☐ U.S. $400–1,199

5. ☐ <U.S. $399

6. ☐ No data

Quality of Life

IN 1990, THE UNITED NATIONS (UN) INTRODUCED THE HUMAN DEVELOPMENT INDEX (HDI) in an attempt to measure how developed each country in the world was. The HDI was intended to replace measuring wealth as the sole determinant of development.

The HDI is a social welfare index that measures three variables:

- **Life expectancy:** which the UN regards as the best measure of the health and safety of a population within a country
- **Education attainment:** in which the mean of years of schooling for adults aged 25 years and expected years of schooling for children of school-going age are both measured
- **Real Gross National Income (GNI) per capita:** in which the GNI is adjusted to the actual purchasing power parity (PPP in U.S. dollars). This calculation reflects what the average income will actually buy in a country

Each variable is given a score ranging from 1.000 (the best) to 0.000 (the poorest). The HDI put Norway at the top in 2010, with a score of 0.939, and Zimbabwe at the bottom, with a score of 0.140.

The world map illustrating HDI shows that:

- Countries with scores exceeding 0.900 correspond closely with More Economically Developed Countries (MEDCs) found in the north which have a high GDP per capita
- Countries with scores below 0.900 equate closely with Less Economically Developed Countries (LEDCs) found in the south
- All of the 10 lowest scoring countries on the HDI are found in Africa

COUNTRIES WITH THE HIGHEST HDI SCORE
(UN 2010)

1.	Norway	0.938
2.	Australia	0.937
3.	New Zealand	0.907
4.	United States	0.902
5.	Ireland	0.895
6.	Liechtenstein	0.891
7.	Netherlands	0.890
8.	Canada	0.888
9.	Sweden	0.885
10.	Germany	0.885

COUNTRIES WITH THE LOWEST HDI SCORE
(UN 2010)

160.	Mali	0.309
161.	Burkina Faso	0.305
162.	Liberia	0.300
163.	Chad	0.295
164.	Guinea-Bissau	0.289
165.	Mozambique	0.284
166.	Burundi	0.282
167.	Niger	0.261
168.	Congo, Dem. Republic	0.239
169.	Zimbabwe	0.140

COLORING KEY

1. ☐ >0.9

2. ☐ 0.8–0.89

3. ☐ 0.7–0.79

4. ☐ 0.6–0.69

5. ☐ 0.5–0.59

6. ☐ <0.49

7. ☐ No data

COLORING TASK

Complete a choropleth map of the world using the number key to highlight the HDI of each country around the world.

THE BRANDT LINE

The Brandt Report, published in 1980 and chaired by the former German Chancellor Willy Brandt, proposed that an imaginary line could be drawn around the world, dividing it into two halves:

- An affluent "North" made up of the richer, more industrialized countries, or the More Economically Developed Countries (MEDCs)

- A poorer "South" consisting of developing, less industrialized countries, or the Less Economically Developed Countries (LEDCs)

The Wretched Dollar

IN THE YEAR 2000, AT THE UNITED NATIONS SUMMIT, eight Millennium Development Goals (MDGs) were agreed to provide a set of development targets for the world to reach by 2015. These were:

- **Goal 1:** Eradicate extreme poverty and hunger
- **Goal 2:** Achieve universal primary education
- **Goal 3:** Promote gender equality and empower women
- **Goal 4:** Reduce child mortality rates
- **Goal 5:** Improve maternal health
- **Goal 6:** Combat HIV/AIDS, malaria, and other diseases
- **Goal 7:** Ensure environmental sustainability
- **Goal 8:** Develop a global partnership for development

Section A of MDG 1 is to halve the proportion of people who live on the equivalent of U.S. $1 a day or less. In 2002, an estimated 17 percent of the world population lived on this amount. They lived on less than or equal to what U.S. $1.08 would have bought in the United States in 1993. In over 20 countries, more than a third of the population lives on less than U.S. $1 a day. All but two of these are in Africa. The largest population living on U.S. $1 a day is in southern Asia, most of whom live in India.

COUNTRIES WITH THE HIGHEST PROPORTIONS OF PEOPLE LIVING ON U.S. $1 A DAY

1. Mali
2. Nigeria
3. Central African Republic
4. Zambia
5. Niger
6. Gambia
7. Burundi
8. Sierra Leone
9. Madagascar
10. Nicaragua

(Based on the percentage of people living on U.S. $1 a day or less adjusted for purchasing power parity)

COLORING KEY

1. <2%

2. 2–5%

3. 6–20%

4. 21–40%

5. 41–60%

6. 61–80%

7. No data

COLORING TASK

Complete a choropleth map of
the world using the number key to
highlight the percentages of people
living on U.S. $1 or less a day of
each country around the world.

HIV Infection Rates

HUMAN IMMUNODEFICIENCY VIRUS (HIV) IS A RETROVIRUS that causes Acquired Immunodeficiency Syndrome (AIDS). HIV slowly invades white blood cells (the body's main defense against infection) so that they fail to operate effectively. This process can take many years (often with few symptoms) until the body's defenses become so weak that infections, disease, and cancer develop. It is this stage that is referred to as AIDS.

HIV is a behavioral disease that is spread through human activity. The most common ways that people become infected with HIV are:

- Having sexual intercourse with an infected partner
- Injecting drugs using a needle or syringe that has been used by someone who is infected
- As a baby of an infected mother, during pregnancy, labor, delivery, or breastfeeding

It is estimated that nearly 39 million people are now living with HIV (the virus that causes AIDS) around the world. This is slightly more than the population of Poland. Another 2.7 million more people become infected with HIV each year. In 2005, the HIV/AIDS pandemic killed more than 2.8 million people. HIV/AIDS is found all over the world, however, some areas are more affected than others. The region that is worst affected is sub-Saharan Africa where in some countries more than one in five adults are infected with HIV. According to the United Nations, the virus is spreading most rapidly in eastern Europe and central Asia, where the number of people living with HIV increased by 67 percent between 2001 and 2008. However, globally, there is some evidence that the number of new cases may now be beginning to stabilize.

10 COUNTRIES WITH THE HIGHEST HIV INFECTION RATES

(2006)

#	Country	Rate
1.	Swaziland	33.4%
2.	Botswana	24.1%
3.	Lesotho	23.2%
4.	Zimbabwe	20.1%
5.	Namibia	19.6%
6.	South Africa	18.8%
7.	Zambia	17.0%
8.	Mozambique	16.1%
9.	Malawi	11.8%
10.	Central African Republic	10.7%

COLORING KEY

1. [ ] No data

2. [ ] 0.0–0.1%

3. [ ] 0.1–0.5%

4. [ ] 0.5–1%

5. [ ] 1–5%

6. [ ] 5–15%

7. [ ] 15–34%

COLORING TASK

Complete a choropleth map of the world using the number key to highlight the HIV infection rate of each country around the world.

How Hungry Are You?

THE GEOGRAPHY OF GLOBAL NUTRITION is a complex issue, but what is clear is that patterns of hunger correlate with patterns of power: those without power tend to suffer first, while those who enjoy access to power seldom experience hunger.

The UN Food and Agricultural Commission (FAO) identified that somewhere in the region of 1.02 billion people suffered from undernutrition worldwide in 2009. The number of undernourished people rose dramatically between 2006 and 2009, and 2009 was a "devastating year for the world's hungry, marking a significant worsening of an already disappointing trend in global food security since 1996." They noted that there were increases in hunger in all of the world's major regions.

On the other hand, affluent consumers everywhere enjoy a diverse and rich diet unprecedented in human history, paying very low prices for both food staples and foods that were once considered "luxury items" (such as oysters, scallops, and shrimps). The result of this has been the emergence of the global obesity pandemic. There has been an increase in global consumption of an energy-dense diet high in fat and low in carbohydrates coupled with a more sedentary lifestyle of motorized transportation, labor-saving devices at home and work, and leisure time dominated by physically undemanding pastimes. The health consequences of obesity can lead to an increased rate in diabetes, high blood pressure, and heart disease.

"Over 852 million people throughout the world do not have enough food to meet their basic nutritional needs."

"As many as a third of the world's people do not meet their physical and intellectual potential because of vitamin and mineral deficiencies."

—UN FAO Committee on World Food Security, May 2005

TOP 5 MOST OBESE COUNTRIES IN THE WORLD

(Percentage of the population that is overweight)

1.	American Samoa	93.5%
2.	Kiribati	81.5%
3.	United States	66.7%
4.	Germany	66.5%
5.	Egypt	66.0%

5 COUNTRIES WITH THE MOST UNDERNOURISHED PEOPLE IN THE WORLD

(Number of undernourished people in millions)

1.	India	217.05
2.	China	154.00
3.	Bangladesh	43.45
4.	Democratic Republic of the Congo	37.00
5.	Pakistan	35.20

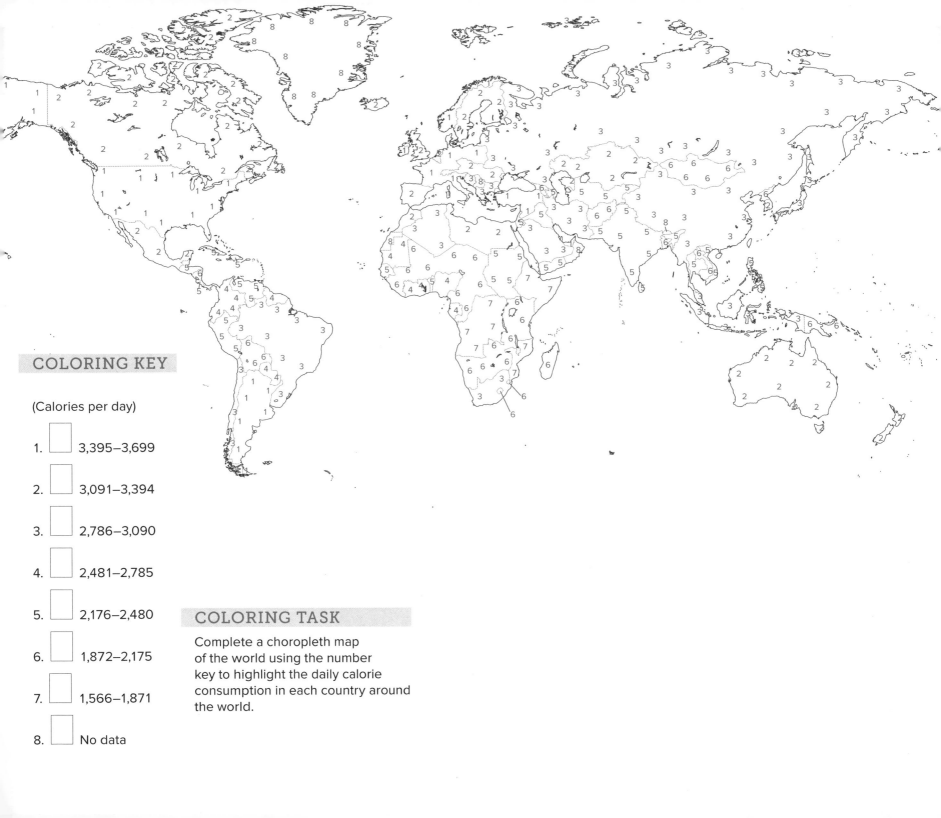

COLORING KEY

(Calories per day)

1. ☐ 3,395–3,699

2. ☐ 3,091–3,394

3. ☐ 2,786–3,090

4. ☐ 2,481–2,785

5. ☐ 2,176–2,480

6. ☐ 1,872–2,175

7. ☐ 1,566–1,871

8. ☐ No data

COLORING TASK

Complete a choropleth map
of the world using the number
key to highlight the daily calorie
consumption in each country around
the world.

Known Oil Reserves

OIL IS USED IN A GREAT MANY WAYS IN MODERN SOCIETY: to fuel our cars, heat buildings, provide electricity, and to make the plastics that we use in everything from milk containers to computers. Most people understand that oil is a finite resource—in other words, it will run out someday. There is only so much oil under the ground and once it is all used up, it will be gone forever. The million dollar question, though, is how much oil is left?

The output of oil has risen significantly from less than a million barrels a day in 1900 to around 83 million barrels today—that's around 3 million barrels fewer than are being used! The International Energy Agency predicts that this demand will rise to 116 million barrels a day by 2030. The map shows which countries have oil reserves and which countries use the most oil. The size of the country shows how much oil the country has, while once colored, the color of the country will indicate how much oil the country uses.

It is difficult for geologists to predict how much oil there is under the ground or how much demand there may be in the future. It is estimated that proven reserves of oil are about 1.12 trillion barrels. However, what is certain is that once "peak" oil is reached, it will become much more difficult and expensive to extract what is left (peak oil is defined as the "maximum rate at which oil can be extracted").

BP estimates that the proven oil reserves are enough to keep the world turning for more than 40 years at today's consumption rates, and the chief economist at BP predicts, "People will run out of demand before they run out of oil." However, the International Energy Authority (IEA) has warned that 64 million more barrels of oil will need to be produced a day to satisfy demand by 2030!

PROVEN OIL RESERVES AT THE END OF 2008

(In million barrels)

1.	Saudi Arabia	264.1
2.	Iran	137.6
3.	Iraq	115.0
4.	Kuwait	101.5
5.	Venezuela	99.4
6.	Russia	79.0
7.	Libya	43.7
8.	Nigeria	36.2
9.	Kazakhstan	39.8
10.	United States	30.5

COLORING KEY

Oil use per day
(in thousands of barrels)

1. >6,000

2.  3,000–5,999

3. 2,000–2,999

4.  1,000–1,999

5.  0–999

COLORING TASK

Complete a choropleth map of
the world using the number key to
highlight the known oil reserves of
each country around the world.

Food Production

EVEN THOUGH THE WORLD'S POPULATION REACHED 6 BILLION IN 1999, doubling from 3 billion in less than 40 years, technological developments in farming have ensured that world food supply has kept pace with this growth:

- On a typical farm in the U.S. in 1830 it took about 300 hours of work to produce 100 bushels of wheat with a walking plow, a harrow, a handful of seed, a sickle, and a flail
- By 2010, the process took just three hours of work to produce the same amount of wheat. The high-tech farm machinery that is available today allows farmers in western countries to cultivate many more acres of land than the simple machines used in the past

The advancement of global food production has ensured that a greater percentage of people are consuming a higher number of calories than they were 40 years ago. World cereal production has doubled since 1970, meat production has tripled since 1961, and the number of fish caught has grown more than six times between 1950 and 1997.

However, with the world's population expected to rise to 9.1 billion by the middle of this century, the United Nations believes that the real challenge for humanity is to increase global food production by 70 percent to ensure that all the people in the world have access to a healthy diet. The UN Development Goal 1 was to halve hunger by 2015, but a new pledge has been put forward—to eradicate hunger by 2025.

TOP 10 WHEAT-PRODUCING COUNTRIES IN 2008

Source: UN FAO

	Country	Production (International $1,000)	Production in Metric Tons (MT)
1.	China	15,805,966	112,463,296
2.	India	11,671,546	78,570,200
3.	United States	9,301,602	68,016,100
4.	Russian Fed.	6,670,506	63,765,140
5.	Canada	4,462,759	28,611,100
6.	France	4,388,762	39,001,700
7.	Pakistan	3,023,994	20,958,800
8.	Australia	2,653,403	21,420,177
9.	Ukraine	2,618,186	25,885,400
10.	Turkey	2,428,920	17,782,000

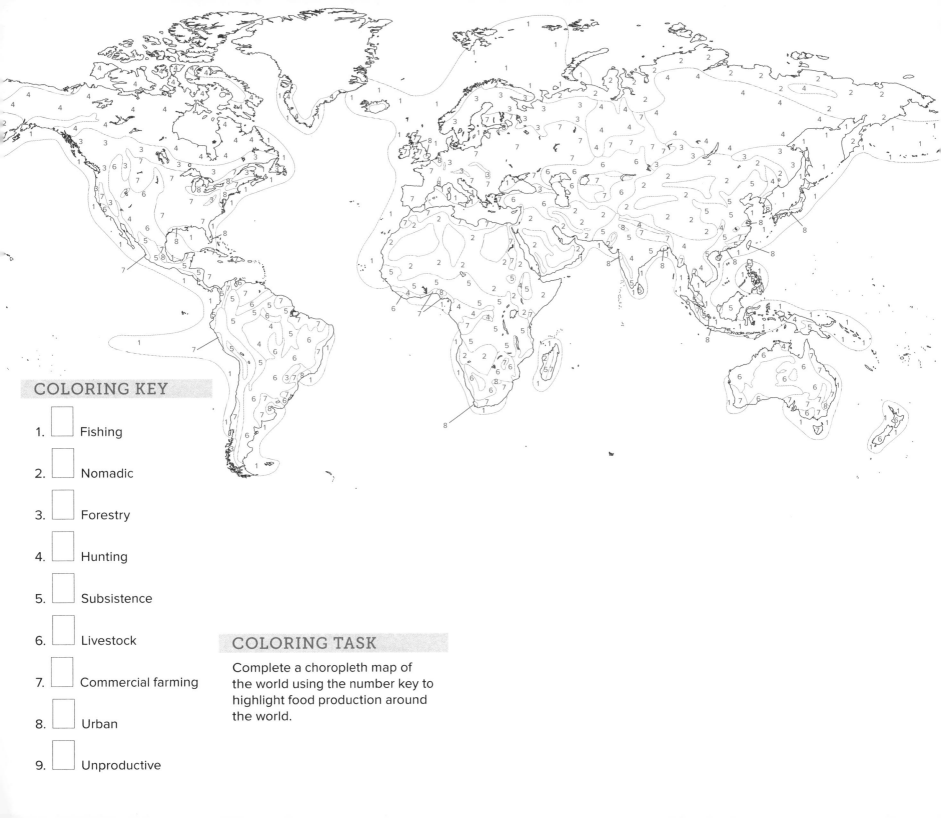

COLORING KEY

1. ☐ Fishing

2. ☐ Nomadic

3. ☐ Forestry

4. ☐ Hunting

5. ☐ Subsistence

6. ☐ Livestock

7. ☐ Commercial farming

8. ☐ Urban

9. ☐ Unproductive

COLORING TASK

Complete a choropleth map of
the world using the number key to
highlight food production around
the world.

The Law of the Sea Treaty

THE LAW OF THE SEA TREATY is formally known as the Third United Nations Convention on the Law of the Sea (UNCLOS III) and was adopted in 1982. The purpose of the agreement is to provide a comprehensive set of rules governing the oceans. The original treaties (UNCLOS I and II) covered four main conventions.

- **The Convention on the Territorial Sea and the Contiguous Zone:** established rights of sovereignty and passage and, Contiguous Zone to extend 12 nautical miles from the baselines

- **The Convention on the High Seas:** access for landlocked nations, concept of "flag state," outlawed transportation of slaves, covered piracy, safety and rescue protocols

- **The Convention on Fishing and Conservation of the Living Resources of the High Seas:** established rights of coastal nations to protect living ocean resources, conservation measures, and measures for dispute resolution

- **The Convention on the Continental Shelf:** established regime governing the superjacent waters, airspace, navigation, fishing, scientific research, and the coastal nation's competence, delimitation, and tunneling

When UNCLOS III came into force in November 1994 several key areas were added:

- **Baselines:** the boundary from which countries can measure their sovereignty
- **Internal waters:** those that are on the landward side of the baseline
- **Territorial sea:** the area of water extending 12 miles from the baseline in which a nation has exclusive sovereignty over the water, seabed, and airspace
- **Exclusive Economic Zone:** a region that stretches 200 miles from the baseline where a nation can exploit raw materials
- **High seas:** waters beyond the Exclusive Economic Zone are considered High seas

10 THINGS TO REMEMBER

1. About 90 percent of world trade is carried by the international shipping industry.

2. Without shipping, intercontinental trade, the bulk transportation of raw materials, and the import/export of affordable food and manufactured goods would simply not be possible.

3. Shipping is regulated globally by the International Maritime Organization (IMO).

4. Shipping is the least environmentally damaging form of commercial transportation.

5. There are around 50,000 merchant ships trading internationally.

6. The world fleet is registered in over 150 nations and manned by over 1 million seafarers of virtually every nationality.

7. The United Nations Conference on Trade and Development (UNCTAD) estimates that the operation of merchant ships contributes about 5 percent of total world trade.

8. Over the last four decades, total seaborne trade has quadrupled, from just over 8 thousand billion tonne-miles in 1968 to over 32 thousand billion tonne-miles in 2008.

9. APM Maersk is the largest container company, with 14.4 percent of the global market.

10. Ships are technically sophisticated, high value assets and larger high-tech vessels can cost over U.S.$200 million to build.

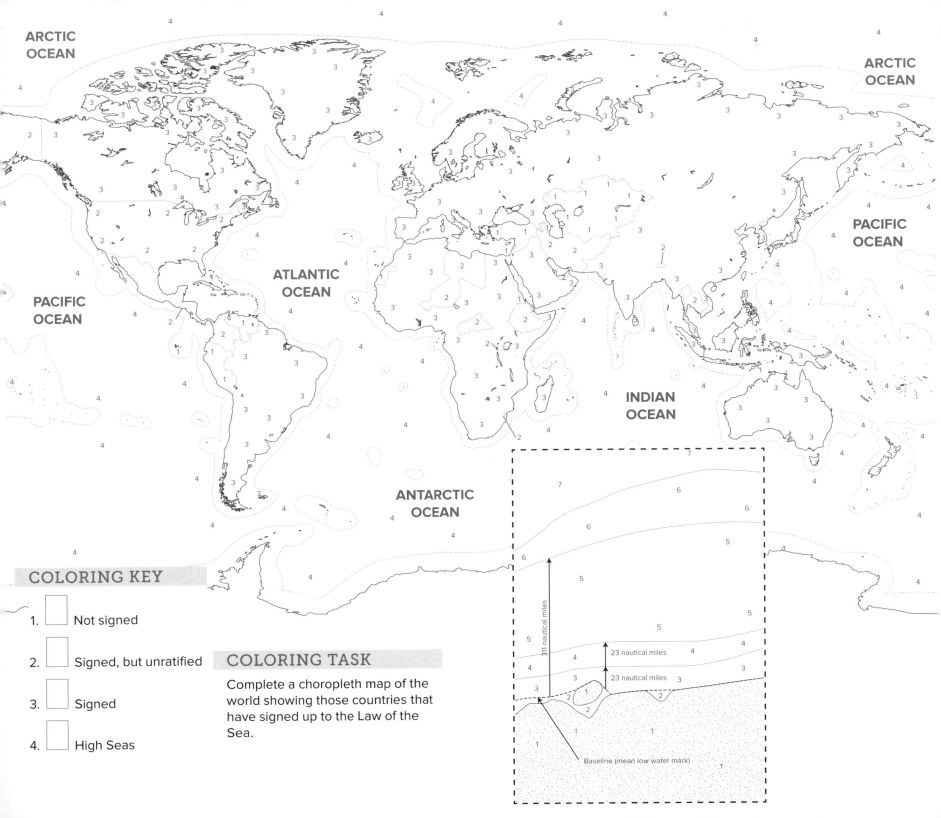

ARCTIC
OCEAN

ARCTIC
OCEAN

PACIFIC
OCEAN

PACIFIC
OCEAN

ATLANTIC
OCEAN

INDIAN
OCEAN

ANTARCTIC
OCEAN

COLORING KEY

1. Not signed

2. Signed, but unratified

3. Signed

4. High Seas

COLORING TASK

Complete a choropleth map of the world showing those countries that have signed up to the Law of the Sea.

311 nautical miles

23 nautical miles

23 nautical miles

Baseline (mean low water mark)

AROUND 40 PERCENT OF THE WORLD'S POPULATION experience water shortages, and this is likely to increase as the world's population continues to grow. Access to safe drinking water will become an increasing problem for people in many of the poorest countries of the world. At present, fresh water is not scarce:

- Around half of the annual global freshwater runoff is available for human use in agriculture, industry, or domestic use
- However, much fresh water is inaccessible or only available at certain times of the year
- Many parts of the world suffer from "water stress." This is where the demand for water exceeds the amount available during a period of time or when the water is not of good enough quality to use. Many lakes, rivers, and groundwater supplies are drying up from overuse, and the outcome is shortage of water supplies—and it is usually the poorest people who suffer the most

Causes of water stress include:

- The impact of global warming and climate change on water supplies
- The continuing growth of the human population
- Increasing industrialization and associated demands for water
- Wasting of water—it is estimated that up to 45 percent of fresh water is wasted through leaking pipes, pollution, or draining away
- "Water wars," or conflicts between countries that seek to control the sections of rivers that pass through their country

It is estimated that 1.4 billion people (around 20 percent of the world's population) do not have access to clean water.

WATER FACTS

1. 2.8 billion people suffer from water shortages. This is expected to rise to 3.8 billion by 2015.

2. One flush of a toilet in the U.S. uses as much water as the average person in an LEDC (Less Economically Developed Country) uses in a day.

3. Over 3½ million people die each year from water-related diseases.

4. The water and sanitation crisis claims more lives through disease than any war has.

5. Poor people often pay 5–10 times more per liter of water than wealthy people living in the same city.

6. An American taking a five-minute shower uses more water than a typical person in a developing country slum uses in a whole day.

7. A bathtub holds 39 gallons (151 L) of water. Somebody living in a slum may only get access to 7 gallons (30 L) for all their daily needs.

8. A child dies every 15 seconds from diseases caused by dirty water.

9. 70 percent of water used by humans is used for agriculture.

10. As much as 45 percent of fresh water is wasted due to leaking pipes.

COLORING KEY

Access to improved
drinking water
(% of population)

1. ☐ 91–100%

2. ☐ 76–90%

3. ☐ 50–75%

4. ☐ <50%

5. ☐ No data

COLORING TASK

Complete a choropleth map of the
world showing the percentage of
people that have access to improved
drinking water in each country.

The Enhanced Greenhouse Effect

CLIMATE CHANGE IS PERHAPS THE BIGGEST CHALLENGE FACING HUMANITY. The British Antarctic Survey (BAS), the Stern Review of 2006, and the Intergovernmental Panel on Climate Change (IPCC) in November 2007 all state that global warming is taking place and that there is clear evidence that it is being caused by human activity.

The Greenhouse Effect: Earth's atmosphere is made up of a number of gases that are vital in supporting life. Carbon dioxide is one of these gases and makes up a tiny percent (around 0.3 percent) of the gases in the atmosphere. Carbon dioxide is important because it helps regulate the temperature of the planet—it helps keep Earth 60.8°F (16°C) warmer than it would be without this gas. This keeps Earth at a temperature that is comfortably warm for supporting life. Without it, Earth would be a frozen wasteland.

The Enhanced Greenhouse Effect: The burning of fossil fuels—for energy, cars etc.—releases carbon dioxide into Earth's atmosphere. This is causing a buildup of more carbon dioxide in the atmosphere, which enhances the natural greenhouse effect and is causing global temperatures to rise. This effect has been given the name "global warming."

Evidence that Global Warming is Taking Place:
- Global temperatures have increased since 1850
- The Arctic ice sheet is melting faster than at any time since records began
- Frozen peat bogs in Siberia are thawing, releasing billions of tons of methane gas into the atmosphere
- The amount of snow melting in Greenland each summer has increased by 30 percent since 1975
- The world's glaciers are melting faster than at any point in the past 5,000 years

10 THINGS THAT YOU CAN DO TO REDUCE GLOBAL WARMING

1. **Reduce, reuse, recycle:** make use of what you already have.
2. **Use less heat and air conditioning:** this will reduce energy demands.
3. **Change your lightbulbs:** use energy-efficient light bulbs.
4. **Drive less and drive smart:** this will reduce your emissions.
5. **Buy energy-efficient products:** as these will use less energy.
6. **Use less hot water:** this will reduce energy consumption.
7. **Use the "off" switch:** rather than the standby button.
8. **Plant a tree:** trees absorb carbon dioxide.
9. **Get a report card from your utility company:** find out just how energy efficient your home is.
10. **Encourage others to conserve.**

COLOR YOURSELF **SMART**
GEOGRAPHY RESOURCES AND THE ENVIRONMENT

Effects of Climate Change

THE MAP ON THE RIGHT SHOWS PREDICTED GLOBAL CHANGES in surface temperature for the year 2100, based on increases in temperature since 1990. The effects of such increases could have dramatic consequences for our planet:

Global Impacts of Climate Change:

1. The Gulf Stream and North Atlantic Drift:

The Gulf Stream is a powerful, warm, and swift Atlantic Ocean current that originates in the Gulf of Mexico and influences the climate of the east coast of North America and the west coast of Europe. The North Atlantic Drift is a powerful warm ocean current that continues along the Gulf Stream northeast where it has a considerable warming influence on the climate of Ireland, Great Britain, and Norway. Climatologists believe that increased global temperatures could switch off the Gulf Stream and the North Atlantic Drift, causing temperatures in western Europe to drop by at least 41°F (5°C), causing winters in Great Britain to be similar to those currently experienced in Germany or Russia.

2. El Niño:

El Niño is a sporadic change in wind patterns that occurs in the mid-south Pacific Ocean. The effects of this wind change can cause short-term changes in the climate of those countries that border the Pacific. In Australia, this can bring drought caused by offshore winds, while in Peru, it can cause floods due to onshore winds. The effects of increased global temperatures are thought to be causing the increased frequency of El Niño and risk intensifying the effects.

10 LOCALIZED IMPACTS OF CLIMATE CHANGE

1. **Arctic sea ice:** researchers believe that the tipping point for the total loss of summer sea ice is imminent.

2. **Greenland ice sheet:** total melting could take 300 years or more, but the tipping point that could see irreversible change might occur within 50 years.

3. **West Antarctic ice sheet:** scientists believe it could unexpectedly collapse if it slips into the sea at its warming edges.

4. **Gulf Stream:** few scientists believe it could be switched off completely this century, but its collapse is a possibility.

5. **El Niño:** the southern Pacific Ocean current may be affected by warmer seas, resulting in far-reaching climate change.

6. **Indian monsoon:** relies on temperature difference between land and sea, which could become imbalanced by pollutants that cause localized cooling.

7. **West African monsoon:** in the past, this monsoon has changed, causing the greening of the Sahara Desert, but in the future it could cause droughts.

8. **Amazon rainforest:** a warmer world and further deforestation may cause a collapse of the rain supporting this ecosystem.

9. **Boreal forests:** cold-adapted trees of Siberia and Canada are dying as temperatures rise.

10. **Change** in any of the nine zones listed above could trigger a disproportionately larger change in the future, and could reach their critical point within this century.

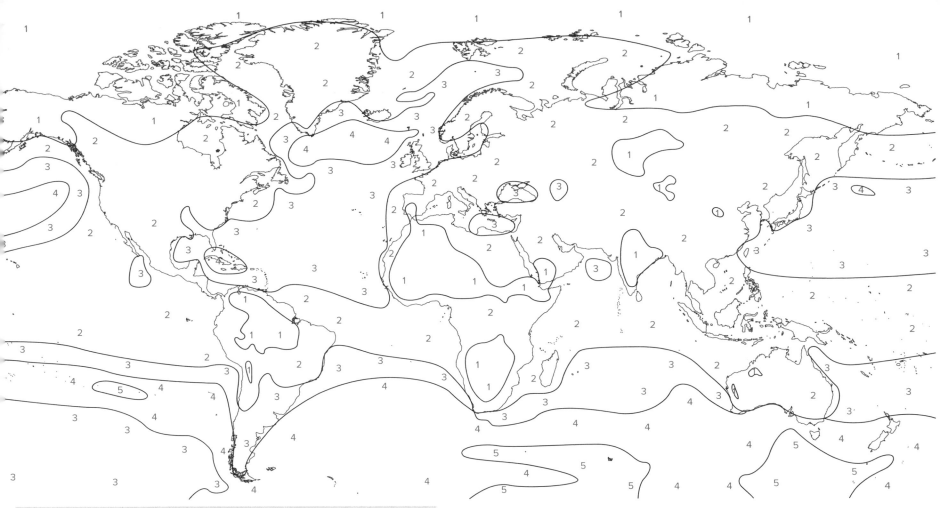

COLORING KEY

Predicted change in temperature (the difference between actual annual average surface temperature, 1969–1990, and predicted annual average surface air temperature, 2070–2100)

1. [ ] 41–50°F (5–10°C) warmer

2. [ ] 37.4–41°F (3–5°C) warmer

3. [ ] 35.6–37.4°F (2–3°C) warmer

4. [ ] 33.8–35.6°F (1–2°C) warmer

5. [ ] 32–35.6°F (0–1°C) warmer

COLORING TASK

Complete a choropleth map of the world showing the predicted increase in annual average surface air temperature, 2070–2100.

Ecological Footprints of Countries

THE TERM "ECOLOGICAL FOOTPRINT" REFERS TO A CALCULATION that estimates the area of Earth's land and water needed to supply resources to an individual or a group of people. It measures the amount of land required to absorb the waste produced by the individual or group.

People are using up Earth's natural resources at least 25 percent faster than the planet can renew them. It takes just nine months for people to use up the resources that the planet requires one year to replenish. Since the 1970s, humanity has been in ecological debt, with annual demand on resources exceeding what Earth can regenerate each year. It is estimated that it now takes one year and six months to regenerate what we use in one year.

One Planet Future

In order to achieve sustainability and ensure that our ecological footprint is reduced to just one planet, action will need to be taken. The "One Planet Future" campaign encourages people to:

- Measure their own ecological footprint
- "Green" their lifestyle—by saving energy and traveling less
- Support sustainable energy solutions—such as wind power
- Consider their diet—think about the distance your food has traveled
- Lobby politicians to take action
- Tackle climate change effectively—Earth's regenerative capacity can no longer keep up with demand, since people are turning resources into waste faster than nature can turn waste back into resources

Source: WWF, *Living Planet Report*, 2006

ECOLOGICAL DEMAND AND SUPPLY IN SELECTED COUNTRIES

Total ecological footprint (in million gha 2003) and per capita ecological footprint (in gha/person)

United States	2,819	9.6
China	2,152	1.6
India	802	0.8
Russia	631	4.4
Japan	556	4.4
Brazil	383	2.1
Germany	375	4.5
France	339	5.6
UK	333	5.6
Mexico	265	2.6

Note: gha = global hectares: the amount of productive land and water required to produce the resources and the waste generated by each country.

Source: WWF, *Living Planet Report*, 2006

COLORING KEY

The number of hectares of land required to produce the resources and absorb the waste generated by an individual.

1. ☐ No data

2. ☐ < 1

3. ☐ 1–2

4. ☐ 2–4

5. ☐ 4–6

6. ☐ >6

COLORING TASK

Complete a choropleth map of the world showing the number of acres of land required by each person in each country.

THE UNITED NATIONS BEGAN AFTER THE SECOND WORLD WAR in 1945, when representatives of 50 countries met in San Francisco to draw up the United Nations Charter. The delegates discussed proposals worked out by the representatives of China, the Soviet Union, the United Kingdom, and the United States in Washington, D.C., from August to October 1944.

The United Nations officially came into existence on October 24, 1945, when the Charter was ratified by China, France, the Soviet Union, the United Kingdom, the United States, and a majority of other signatories. United Nations Day is celebrated on October 24 each year.

There are currently 192 member states of the United Nations, including all fully recognized independent states. The Vatican City is not a member, but it is a permanent observer of proceedings.

The United Nations General Assembly: The General Assembly is the main deliberative organ of the UN. Decisions on important questions, such as those on peace and security, admission of new members, and budgetary matters, require a two-thirds majority. Decisions on other questions are by simple majority.

The United Nations Security Council: The Security Council has responsibility to deal with complaints concerning threats to international peace and security. It is organized to function continuously, and a representative of each of its members must be present at all times at the United Nations Headquarters. There are 15 members of the Security Council, consisting of five veto-wielding permanent members (China, Russia, France, the United Kingdom, and the United States) and 10 elected nonpermanent members with two-year terms.

NONELECTED PERMANENT MEMBERS OF THE UN SECURITY COUNCIL IN 2010

(Parentheses indicate year of term end)

1. Austria (2010)
2. Bosnia and Herzegovina (2011)
3. Brazil (2011)
4. Gabon (2011)
5. Japan (2010)
6. Lebanon (2011)
7. Mexico (2010)
8. Nigeria (2011)
9. Turkey (2010)
10. Uganda (2010)

COLORING KEY

1. ☐ Member of the United Nations

2. ☐ Permanent Member of the UN Security Council

3. ☐ Nonpermanent Member of the UN Security Council

4. ☐ Nonmember of the United Nations

COLORING TASK

Complete a choropleth map of the world showing membership of the United Nations.

The G20

THE GROUP OF 20 (OR, AS IT IS MORE COMMONLY KNOWN, "THE G20") is a forum for international economic cooperation that was formed in 1999. The G20 is a group of 19 countries and the European Union. Members of the G20 countries meet regularly to discuss key global economic and financial issues. The G20 members account for 85 percent of the entire world's economic output, about 80 percent of global trade, and make up more than two-thirds of the world's population.

The G20 represents a range of geographical regions of the world, but has limited membership in order to be able to make effective decisions in short time frames. The G20 works closely with a number of key international institutions, including the International Monetary Fund (IMF), World Bank, World Trade Organization (WTO), Financial Stability Board (FSB), International Labor Organization (ILO), and the Organization for Economic Cooperation and Development (OECD). Representatives from the UN as well as regional organizations such as ASEAN, APEC, and the African Union have attended past G20 Leaders' Summits.

A number of countries have previously been invited to past G20 summits as guests. These guests are selected by the host country. For the 2010 summits, both Canada and South Korea invited Ethiopia (chair of NEPAD), Malawi (chair of the African Union), Vietnam (chair of ASEAN), and Spain. Canada also invited the Netherlands (the world's sixteenth largest economy), while Korea invited Singapore.

G20 members

Argentina	Japan
Australia	Republic of Korea
Brazil	Mexico
Canada	Russia
China	Saudi Arabia
France	South Africa
Germany	Turkey
Indonesia	United Kingdom
India	United States
Italy	European Union

THE 10 MOST RECENT G20 MEETINGS FOR FINANCE MINISTERS

2001	Ottawa, Canada
2002	New Delhi, India
2003	Morelia, Mexico
2004	Berlin, Germany
2005	Beijing, China
2006	Melbourne, Australia
2007	Cape Town, South Africa
2008	São Paulo, Brazil
2009	Horsham, Australia (March)
	London, England (September)
	St. Andrews, Scotland (November)
2010	Incheon, South Korea (February)
	Toronto, Canada (June)
	Seoul, South Korea (November)

St. Andrews, 200■
London & Horsham, 200■
Ottawa, 2001
Toronto, 2010
Morelia, 2003 ■
São Paolo, 2008 ■
Cape Town, 2007 ■
Melbourne, 2006
Berlin, 2004
New Delhi, 2002 ■
Beijing, 2005 ■
Seoul & Incheon, 2010

COLORING KEY

1.  Member states

2. [ ] Represented by the EU

COLORING TASK

Complete a choropleth map of
the world showing membership
of the G20.

Trading Blocs

THERE ARE A NUMBER OF TRADING BLOCS AROUND THE WORLD. A trading bloc is a group of countries that have joined together to improve their economic interests and trading patterns. Some of the main players include the EU, NAFTA, ASEAN, MERCOSUR, and OPEC.

The countries in these blocs group together in geographical areas and encourage trade within the bloc by removing duties on goods for members and creating barriers to outsiders by placing tariffs on goods from outside. The benefits of these trading blocs have been great for some countries—for example, some of Asia's NIC (Newly Industrializing Countries) such as Thailand have seen huge economic growth.

Key Players in World Trade

- **WTO** (World Trade Organization) was formed in 1993 with the aim of cutting trade barriers that stop countries from trading freely with each other, so that goods can flow more easily
- **OECD** (Organization for Economic Cooperation and Development) is a global "think tank" for the world's 30 richest countries
- **OPEC** (Organization of Petroleum Exporting Countries) is a group of 11 countries who supply 40 percent of the world's oil and work together to regulate the global oil market and stabilize prices
- **World Bank** is based in Washington, D.C. and exists to promote global investment and provide loans to countries for development projects
- **IMF** (International Monetary Fund) provides loans to governments of countries that face financial difficulty. The loans are often tied to "structural re-adjustment packages" (SAPs) which usually involve governments being forced to sell off any assets that they may have

10 KEY TRADING BLOCS AROUND THE WORLD*

1. **ASEAN:** Association of South East Asian Nations
2. **NAFTA:** North American Free Trade Agreement
3. **MERCOSUR:** Mercado Común del Sur (Southern Common Market)
4. **OPEC:** Organization of Petroleum Exporting Countries
5. **OECD:** Organization of Economic Co-operation and Development
6. **APEC:** Asia Pacific Economic Co-operation
7. **LAIA:** Latin America Integration Association
8. **EU:** European Union
9. **AU:** African Union
10. **SAARC:** South Asian Association for Regional Co-operation

COLORING KEY

Membership of five key trading blocs

1. ☐ ASEAN

2. ☐ NAFTA

3. ☐ MERCOSUR

4. ☐ MERCOSUR (awaiting ratification)

5. ☐ MERCOSUR associates

COLORING TASK

Complete a choropleth map of the world showing membership of the five key trading blocs.

The European Union

THE EUROPEAN UNION (EU) IS AN ECONOMIC AND POLITICAL UNION of 27 countries located within Europe. With almost 500 million citizens, the EU countries combined generate an estimated 30 percent share of the world's wealth, worth $18.4 trillion. The EU traces its origins to the European Coal and Steel Community formed among six countries in 1951 and the Treaty of Rome in 1957. Originally known as the European Economic Community (EEC), it became known as the European Union in 1992 after the signing of the Maastrict Treaty. Since then, the EU has grown in size through the accession of new countries, and new policy areas have been added to the remit of the EU's institutions.

A common currency, the Euro (€), has been adopted by 16 member states that comprise the "Eurozone." The UK has opted not to adopt the Euro as a common currency, and Denmark is holding a referendum on whether their people wish to have the Euro as their currency.

The other eight members are all committed to join the Euro. The European Parliament is elected every five years by the member states' citizens. The European Parliament meets in Brussels in Belgium and Strasbourg in France. The EU has common policies on trade, agriculture, fisheries, and regional development. The EU has developed a single market through the same system of laws which apply in all member states, ensuring the freedom of movement for people, goods, services, and capital.

Currently, six countries have applied to become members of the European Union. They are Turkey, Croatia, Macedonia, Montenegro, Albania, and Iceland. Norway's government has applied three times to become a member, but on all three occasions, their electorate have voted against allowing the country to join. Possible future members could also include Ukraine and Serbia.

FORMATION OF THE EUROPEAN UNION

The EU has grown over the last half century. The periods of expansion are as follows:

1949: Creation of Council of Europe

1951: Schuman Plan in which six countries sign a treaty to run their heavy industries under common agreement

1957: Treaties of Rome are signed, forming the European Economic Community (EEC)

1957 Initial members of the EEC: Belgium, France, Italy, Luxembourg, the Netherlands, and West Germany

1967: EEC became the EC

1973: Denmark, Ireland, and United Kingdom join the EC

1981: Greece joins the EC

1985: Portugal and Spain join the EC

1993: EC became the EU (European Union)

1995: Austria, Finland, and Sweden join the EU

2004: Cyprus, Czech Republic, Estonia, Hungary, Latvia, Lithuania, Malta, Slovakia, Slovenia, and Poland join the EU

2007: Bulgaria and Romania join the EU

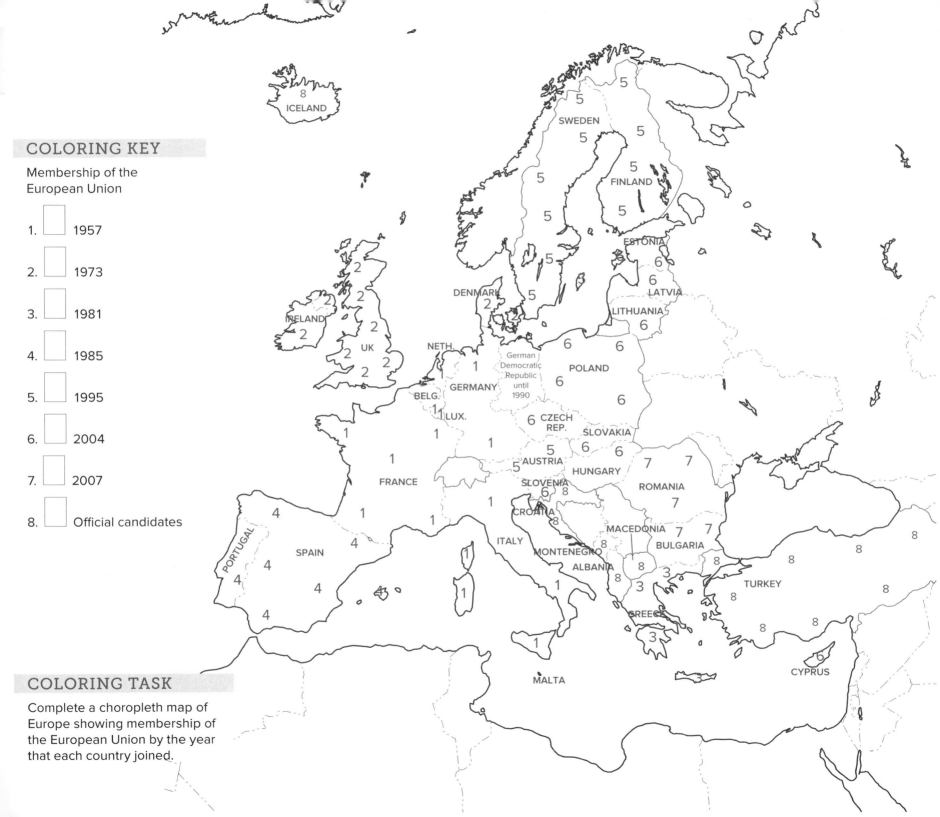

COLORING KEY

Membership of the
European Union

1. ☐ 1957

2. ☐ 1973

3. ☐ 1981

4. ☐ 1985

5. ☐ 1995

6. ☐ 2004

7. ☐ 2007

8. ☐ Official candidates

COLORING TASK

Complete a choropleth map of
Europe showing membership of
the European Union by the year
that each country joined.

Brazil

BRAZIL IS THE SIXTH-MOST POPULATED COUNTRY IN THE WORLD and home to over half of South America's population. Over 80 percent of Brazil's population live in urban areas, many in the large cities found in the southeastern corner of Brazil, such as Rio de Janeiro, São Paulo, and Belo Horizonte. It is South America's most influential country, an economic powerhouse, and one of the world's largest democracies.

More than one-third of Brazil is made up of tropical rainforest. The Amazon River winds its way through the rainforests from its source high up in the Andes, across Brazil, and out into the Atlantic Ocean. It is the world's second-largest river and contains more than 20 percent of the world's fresh water. Large areas of the Amazon forests have been cleared by deforestation, and in 2005 the Brazilian government reported that one-fifth of the rainforest had been cleared. The government has since made efforts to control illegal logging and introduce improved certification of land ownership in order to bring the deforestation under control.

Brazil is rich in mineral deposits and resources, especially iron ore. There has been large-scale development of offshore oil fields, which has enabled Brazil to become self-sufficient in oil. Economic growth in Brazil has been erratic, with periods of boom and bust and periods of high inflation. In 2002, Brazil had to turn to the IMF for financial support to avoid defaulting on loans. However, by 2005, strong economic growth enabled Brazil to pay off its entire $15 billion loan two years early.

Industry now accounts for 39 percent of Brazil's GDP. The recent resurgence in economic development has seen Brazil move into a position in which the country is poised to become a future superpower.

10 FACTS ABOUT BRAZIL

1. **Full name:** Federative Republic of Brazil
2. **Population:** 195.4 million (UN, 2010)
3. **Capital city:** Brasilia
4. **Largest city:** São Paulo
5. **Area:** 3.3 million sq miles (8.55 million km²)
6. **Major language:** Portuguese
7. **Major religion:** Christianity
8. **Life expectancy:** 70 years (men), 77 years (women) (UN)
9. **Main exports:** Manufactured goods, iron ore, coffee, oranges, other agricultural produce
10. **GNI per capita:** U.S. $8,040 (World Bank, 2009)

DID YOU KNOW?
THE BRICs

The BRICs is an acronym for Brazil, Russia, India, and China. These four countries are significant, as they are on the brink of becoming world superpowers.

VENEZUELA

COLUMBIA

SURINAME

French Guiana (FRANCE)

NORTH ATLANTIC OCEAN

Boa Vista

GUYANA

AMAPA

Olapoque

PICO DA NEBLINA ▲

GUIANA HIGHLANDS

Macapa

Rio Negro

Amazon

Belem

Manaus

Santarem

Sao Luis

Parnaiba

Careiro

Altamira

Fortaleza

Itaituba

Maraba

MARANHAO

Teresina

CEARA

RIO GRANDE DO NORTE

AMAZONAS

PARA

Carajas

Araguaina

Picos

Natal

Benjamin Constant

PIAUI

PARAIBA

Joao Pessoa

SERRA DO CACHIMBO

Rio Xingu

Araguaia

Salgueiro

PERNAMBUCO

Recife

Cruzeiro do Sul

Rio Purus

Humaita

Rio Teles Pires

Cachimbo

SERRA DOS GRADAUS

SERRA DA CAATINGA

Juazeiro

ALAGOAS

Maceio

BRAZIL

Rio Madeira

Porto Velho

SERRA FORMOSA

Palmas

TOCANTINS

Tocantins

CHAPADA DIAMANTINA

Aracaju

SERGIPE

ACRE

Rio Branco

Boca do Acre

Guajara-Mirim

RONDONIA

Barreiras

BAHIA

PERU

Assis Brasil

CHAPADA DOS PARECIS

Rio Juruena

MATO GROSSO

Rio Araguaia

GOIAS

SERRA GERAI DE GAIAS

Rio Sao Francisco

Vitoria da Conquista

Ilheus

Salvador

BOLIVIA

PLANALTO DE MATO GROSSO

Cuiaba

Rondonopolis

Brasilia ★

DISTRITO FEDERAL

SERRA DO ESPINHACO

Caceres

Goiania

MINAS GERAIS

ESPIRITO SANTO

Corumba

Rio Paranaiba

Uberlandia

SOUTH PACIFIC OCEAN

Campo Grande

Santa Fe do Sul

Rio Grande

MATO GROSSO DO SUL

Panorama

SAO PAULO

Belo Horizonte

Vitoria

Ponta Pora

Rio Paraguay

PARAGUAY

Sao Paulo

RIO DE JANEIRO

PARANA

Santos

Rio de Janeiro

Curitiba

SOUTH ATLANTIC OCEAN

Foz do Iguacu

Sao Francisco do Sul

Rio Parana

SANTA CATARINA

Florianopolis

CHILE

RIO GRANDE DO SUL

Santa Maria

Porto Alegre

ARGENTINA

Rio Parana

URUGUAY

Rio Grande

Russia

RUSSIA EMERGED FROM THE ASHES OF THE COLLAPSE OF THE FORMER SOVIET UNION IN 1991. Russia was the largest of the 15 states that made up the USSR, and it is now emerging from a decade of economic and political turmoil to reassert itself as a potential superpower in its own right.

Russia has the world's largest reserves of natural gas and is the second-largest producer of oil after Saudi Arabia. These two natural resources accounted for over 60 percent of Russia's exports in 2007 and have been central to Russia's economic recovery after the near collapse of its economy in 1998.

Gazprom is Russia's largest energy company, and the world's largest producer and exporter of gas, and it supplies a growing share of Europe's needs. The company is 50 percent owned by the state, while the other 50 percent is owned by shareholders—some of whom have made huge personal fortunes from the company. Gazprom controls one third of the world's natural gas reserves, and its profits have enabled the Russian government to repay IMF loans that were accrued in the 1990s.

The new economic strength that Russia has recently gained allowed Vladimir Putin to enhance state control over political institutions and the media. He has been a popular figure in Russian politics and received extensive support for his policies as prime minister, president, and now prime minister again. Despite its outdated, Cold War-era military hardware, Russia still has one of the world's largest armies.

The vast Russian landmass covers more than 6.6 million sq miles (17 million km²), with a climate ranging from the Arctic north to a generally more temperate south. The country spans nine time zones and is the largest country on Earth in terms of surface area. Large parts of the country, particularly in the north and east, are inhospitable and sparsely populated.

10 FACTS ABOUT RUSSIA

1. **Full name:** Russian Federation
2. **Population:** 140.3 million (UN, 2010)
3. **Capital:** Moscow
4. **Area:** 6.6 million sq miles (17 million km²)
5. **Major language:** Russian
6. **Major religions:** Christianity, Islam
7. **Life expectancy:** 62 years (men), 74 years (women) (UN)
8. **Monetary unit:** 1 rouble = 100 kopecks
9. **Main exports:** Oil and oil products, natural gas, wood, and wood products, metals, chemicals, weapons, military equipment
10. **GNI per capita:** U.S. $9,370 (World Bank, 2009)

COLORING TASK

Color the regions of Russia as well as the main physical features (such as rivers and mountains) in the map above.

India

INDIA IS THE WORLD'S LARGEST DEMOCRACY and has the second-largest population on Earth, with over 1 billion inhabitants. It has emerged since the 1990s as a major economic powerhouse. It has a major cultural influence both in South Asia and beyond as well as a fast-growing and powerful economy.

India is today the fourth-largest economy in the world and is a leading world exporter, with an annual growth rate of 8 percent, and has seen GNP per capita increase from U.S. $300 in 1991 to U.S. $950 in 2009. The manufacturing industry has become increasingly important to the Indian economy and has seen a major transformation since 1991. Many international firms are rushing to invest in India attracted by cheap labor and an emerging consumer market. An increasing number of Indians can now afford to buy Western goods, creating a huge consumer market with an insatiable appetite.

India has a federal style of government similar to that of the U.S., with the government defining national policy and direction. A second tier of 28 states, also with elected legislatures, has responsibility for other areas. While India is held up as a beacon of democracy, it has led to burdensome bureaucracy and slow rates of development in some parts of the country. Communal, caste, and regional tensions cause friction in Indian politics, sometimes threatening its long-standing democratic and secular ethos.

Today, India still struggles to address extreme poverty and to develop its rural population. There is a stark contrast between the rapid growth of national and global technology-based industries and the very slow development in traditional and rural economies. As it heads toward superpower status, with its rapidly expanding economy and massive population, India is still tackling huge social, economic, and environmental problems.

10 FACTS ABOUT INDIA

1. **Full name:** Republic of India
2. **Population:** 1.2 billion (UN, 2010)
3. **Capital:** New Delhi
4. **Most-populated city:** Mumbai (Bombay)
5. **Area:** 1.2 million sq miles (3.1 million km²), excluding Indian-administered Kashmir (38,830 sq miles/100,569 sq km²)
6. **Major languages:** Hindi, English, and at least 16 other official languages
7. **Major religions:** Hinduism, Islam, Christianity, Sikhism, Buddhism, Jainism
8. **Life expectancy:** 64 years (men), 67 years (women) (UN)
9. **Main exports:** Agricultural products, textile goods, gems and jewelry, software services and technology, engineering goods, chemicals, leather products
10. **GNI per capita:** U.S. $1,180 (World Bank, 2009)

AFGHANISTAN

CHINA

PAKISTAN

Srinagar · Kargil
Leh
JAMMU AND KASHMIR

Jammu
Pathankot
HIMACHAL PRADESH
Amritsar
Sutlej · Simla
PUNJAB Chandigarh
Dehra Dun
CHANDIGARH **HARYANA** **DELHI** **UTTAR-ANCHAL**
Delhi
New Delhi · Bareilly
Bikaner
RAJASTHAN **UTTAR PRADESH**
Jaisalmer
Agra
Jodhpur **GREAT INDIAN DESERT** Jaipur Kanpur
Gwalior
Udaipur Kota Allahabad Benares
Gandhinagar **INDIA**
Kandla
Okha Ahmadabad **Bhopal** · Jabalpur
Jamnagar Indore
Vadodara **MADHYA PRADESH**
GUJARAT Narmada
Diu Surat
Daman Silvassa
DAMAN AND DIU **DADRA AND NAGAR HAVELI**
Mumbai (Bombay)
· Pune

NEPAL **GREAT HIMALAYA RANGE**
SIKKIM
Gangtok **BHUTAN**
Dibrugarh Tinsukia
Lucknow
Gorakhpur **NAGALAND**
Ghaghara Dispur **ASSAM** Kohima
Ganges Shillong **MANIPUR**
Patna **MEGHALAYA** Silchar
BIHAR Ganges **BANGLADESH** Imphal
JHARKHAND **TRIPURA** Agartala
Ranchi **WEST BENGAL** Aizawl
CHHATTISGARH Jamshedpur Kolkata **MIZORAM**
Calcutta
Haldia **MYANMAR**
Raipur Balasore Mouths of the Ganges
Nagpur
MAHARASHTRA Mahanadi Cuttack
CRISSA Paradip
Godavari Bhubaneswar **Bay of Bengal**

Arabian Sea

· Hyderabad Vishakhapatnam
Krishna Kakinada
ANDHRA PRADESH
PONDICHERRY

Panaji
Marmagao **KARNATAKA**
GOA Guntakal
EASTERN GHATS

ANDAMAN ISLANDS · Port Blair

Andaman Sea

ANDAMAN AND NICOBAR ISLANDS

Mangalore
Bangalore
PONDICHERRY Chennai (Madras)
Coleroon Pondicherry
Calicut Coimbatore **PONDICHERRY**
Kavaratti Cuddalore
WESTERN GHATS **PONDICHERRY**
LAKSHADWEEP **TAMIL NADU**
KERALA Cochin Madurai
Palk Strait
Trivandrum Tuticorin
Gulf of Mannar
SRI LANKA

NICOBAR ISLANDS

Laccadive Sea

MALDIVES **INDIAN OCEAN** **INDONESIA**

COLORING TASK

Color the regions of India as well as the main physical features (such as rivers and mountains) in the map above.

China

CHINA IS THE WORLD'S MOST POPULATED COUNTRY, with individuals numbering well over 1.3 billion people. It has a long and turbulent history with a continuous culture stretching back nearly 4,000 years. Historical events and political ideas have had a huge influence on the current politics and economy of the country.

China's economy has undergone massive changes since the 1980s, with its economy now doubling in size every eight years. China now has the world's fastest-growing economy and is undergoing what has been described as a second industrial revolution. It has sustained the largest GDP growth in history. Public spending on health and education over the past 50 years has ensured that the country has a healthy, literate, and skilled workforce.

China's integration into the global economy has been a key strategy in its modernization process. The government has encouraged companies to develop and compete with global firms. China joined the World Trade Organization (WTO) in 2001, which has been regarded by many as a positive development that has benefitted the world economy.

China's huge economic growth has seen the demand for raw materials increase on a massive scale and has accounted for 90 percent of all sea traffic in the first decade of this century. The Chinese steel industry produces 500 million metric tons of steel each year—about one-third of the world's output, and four times larger than that of the U.S. It is now by far the largest producer and consumer of steel in the world, having overtaken the U.S. in 2001.

Many of the elements that make up the foundation of the modern world originated in China, including paper, gunpowder, credit banking, the compass, and paper money. China aspires to superpower status. It has a formidable military might and a rapidly advancing economy.

10 FACTS ABOUT CHINA

1. **Full name:** People's Republic of China
2. **Population:** 1.35 billion (UN, 2010)
3. **Capital:** Beijing
4. **Largest city:** Shanghai
5. **Area:** 3.7 million sq miles (9.6 million km²)
6. **Major language:** Mandarin Chinese
7. **Major religions:** Buddhism, Christianity, Islam, Taoism
8. **Life expectancy:** 72 years (men), 76 years (women) (UN)
9. **Main exports:** Manufactured goods, including textiles, garments, electronics, arms
10. **GNI per capita:** U.S. $3,590 (World Bank, 2009)

KAZAKHSTAN

RUSSIA

KYRGYZSTAN

Irtysh

Karamay

TIAN SHAN

Urumqi ⊙

HEILONGJIANG

Amur

Qiqihar

Jixi

Harbin

JILIN

Jilin

Changchung

MONGOLIA

Sea of
Japan
(East Sea)

TAKLIMAN DESERT

Korla

Hami

GOBI DESERT

NEI MONGOL

Kashi

XINJIANG

Fuxin

Shenyang

Benxi

LIAONING

NORTH
KOREA

ISTAN

Hotan

Qiemo

Hohhot ⊙

Yumen

Baotou

BEIJING
Beijing ★ TIANJIN
Tianjin

Dalian

SOUTH
KOREA

Golmud

QINGHAI

Yinchuan ⊙

NINGXIA

Xining ⊙

Lanzhou ⊙

Taiyuan ⊙

SHANXI

Huang

HEBEI

⊙ Shijiazhuang

Jinan ⊙

Taian

Qingdao

Yellow
Sea

JAPAN

CHINA

GANSU

Xian ⊙

Zhengzhou ⊙

SHANDONG

Kaifeng

JIANGSU

HENAN

XIZANG

PLATEAU OF TIBET

Mianyang

SHAANXI

Nanjing ⊙

Hefei ⊙

HUBEI

ANHUI

Shanghai
SHANGHAI

NEPAL

Lhasa ⊙

Chengdu ⊙

SICHUAN

Zigong

Yangtze

Chongqing

Wuhan ⊙

Yueyang

Hangzhou ⊙

ZHEJIANG

Nanchang ⊙

Ningbo

Wenzhou

East China
Sea

BHUTAN

Dukou

Changsha ⊙

GUIZHOU

HUNAN

JIANGXI

Fuzhou

FUJIAN

Taipei ⊙

INDIA

BANGLADESH

⊙ Guiyang

Kunming ⊙

YUNNAN

Xiamen

GUANGXI

GUANGDONG

Guangzhou ⊙

Shantou

Kao-Hsiung

T'ai-chung

T'ai-nan
Taiwan

COLORING TASK

Color the regions of China as well
as the main physical features (such
as rivers and mountains) in the
map above.

MYANMAR

Nanning ⊙

VIETNAM

LAOS

Haikou

HAINAN

Hong Kong
HONG KONG

South China Sea

PHILIPPINES

THAILAND

Countries at War

MANY PARTS OF THE WORLD HAVE SEEN ARMED CONFLICT SINCE 1990. Just a few examples of these conflicts include:

Wars in the former Yugoslavia, 1991–1995: The former Socialist Federal Republic of Yugoslavia began to break up in 1991, with a series of armed conflicts between the constituent states. Conflict began when ethnic Serbs living in the Krajina area of Croatia attempted independence from Yugoslavia in early 1991. It was not until June 1991 that Slovenia and Croatia began to break away, and this led to a military attack by the Serb controlled Federal Army and Air Force. Further fighting broke out in Bosnia after it, too, declared independence. Eventually, after four years of conflict, a peace agreement was signed in November 1995.

Somali Civil War 1991: The Somali Civil War began in 1991 and has continued intermittently until the present day. The conflict began when President Siad Barre was ousted from power in January 1991. This was followed by a counter-revolution to reinstate him as leader. Following armed conflict across the country, the Somililand region declared itself independent (although it still has no official recognition as an independent state). UN Peacekeepers arrived in 1993 to try and provide humanitarian aid, but withdrew from the region in 1995. Somalia still remains a mainly lawless region without order.

War in Darfur, Sudan, 2003–2009: This was a civil war that began in 1993 when the Sudan Liberation Army (SLA) and the Justice and Equality Movement (JEM)—both based in the Darfur area of Sudan—took up arms against the Sudanese government. The war is thought to have caused over 300,000 deaths. There are still ongoing issues with local banditry in the country, but large-scale conflict has now come to an end.

10 MAJOR CONFLICTS SINCE 1990

1. 1991   War in former Yugoslavia
2. 1991   Somali Civil War
3. 1996   War in Lebanon
4. 1997   Republic of the Congo Civil War
5. 1998   Eritrea–Ethiopian War
6. 1998   Second Congo War
7. 2001   War in Afghanistan
8. 2003   War in Darfur
9. 2003   Iraq War
10. 2006   Lebanon War

Russia 1

UK

Slovenia 1 Croatia 1
Bosnia 1 Serbia 1 Georgia 1
1 Turkey 1 Azerbaijan 1 Afghanistan 1
Lebanon 1 Iraq 1 Iran 1 Pakistan 1 Nepal 1 India 1
Israel 1 Qatar 1 India 1
Algeria 1 Egypt 1 Burma (Myanmar) 1

Haiti 1

Senegal 1 Chad 1 Yemen 1
Guinea 1 Nigeria 1 Sudan 1 Cambodia 1
Sierra Leonne 1 1 Ethiopia 1
Colombia 1 Ivory Coast 1 Uganda 1 Somalia 1 Indonesia 1
Liberia 1 Congo 1 Kenya 1 Indonesia 1
Peru 1 Democratic Republic of Congo 1 Rwanda 1 Papua New Guinea 1
Angola 1 Burundi 1

South Africa

COLORING TASK

Complete a choropleth map to
show countries at war since 1990.

Israel and the Middle East

THE ISRAEL-PALESTINE CONFLICT DATES BACK MANY YEARS, but the failed attempt to establish two states in 1948 has been the cause of the current dispute over land rights in the region. There is still no official state of Palestine.

History of the Conflict

1920–1930: After World War I, the Ottoman Empire was split up, and Palestine came under British control. Increasing anti-Semitimism in Europe led to more than 75,000 Jews migrating to the area.

1947: Great Britain relinquishes its mandate over Palestine, and responsibility is handed to the United Nations. A UN proposal is put forward to divide the land into separate Jewish and Arab states. The Arabs reject the plan.

1948–1949: The new state of Israel is formed. Neighboring Arab countries invade but are driven back. Many Palestinian Arabs flee to nearby countries as Israel moves forward and claims more territory. A new armistice line is agreed, but there is no official peace agreement.

1967—The Six Day War: Israel attacked Arab troops on its borders and took control of more land—the Golan Heights from Syria, Sinai and Gaza from Egypt, and the West Bank and the Old City of Jerusalem from Jordan. Israel still retains these land areas, and peace talks have centered around returning to pre-1967 borders.

Peace Efforts in the Region

Despite the Oslo Accords, violence has continued in the region. There have been suicide bombings and attacks by militant Palestinian groups and assassinations and blockades by the Israeli Army. Israel also went to war with Lebanon for a second time in 2006. Peace talks have continued in 2010, but there has still been no agreement between Israel and Palestine.

5 FACTS ABOUT ISRAEL

1. **Full name:** State of Israel
2. **Population:** 7.3 million (UN, 2010)
3. **Major languages:** Hebrew, Arabic
4. **Life expectancy:** 79 years (men), 83 years (women) (UN)
5. **GNI per capita (Israel only):** U.S. $25,740 (World Bank, 2009)

5 FACTS ABOUT PALESTINE

1. **Population:** 4.4 million (UN, 2010)
2. **Area:** Palestinian Ministry of Information cites 2,305 sq miles (5,970 km²) for West Bank territories and 141 sq miles (365 km²) for Gaza
3. **Major language:** Arabic
4. **Life expectancy:** 73 years (men), 76 years (women) (UN)
5. **GNI per capita:** U.S. $1,230 (estimated, World Bank, 2007)

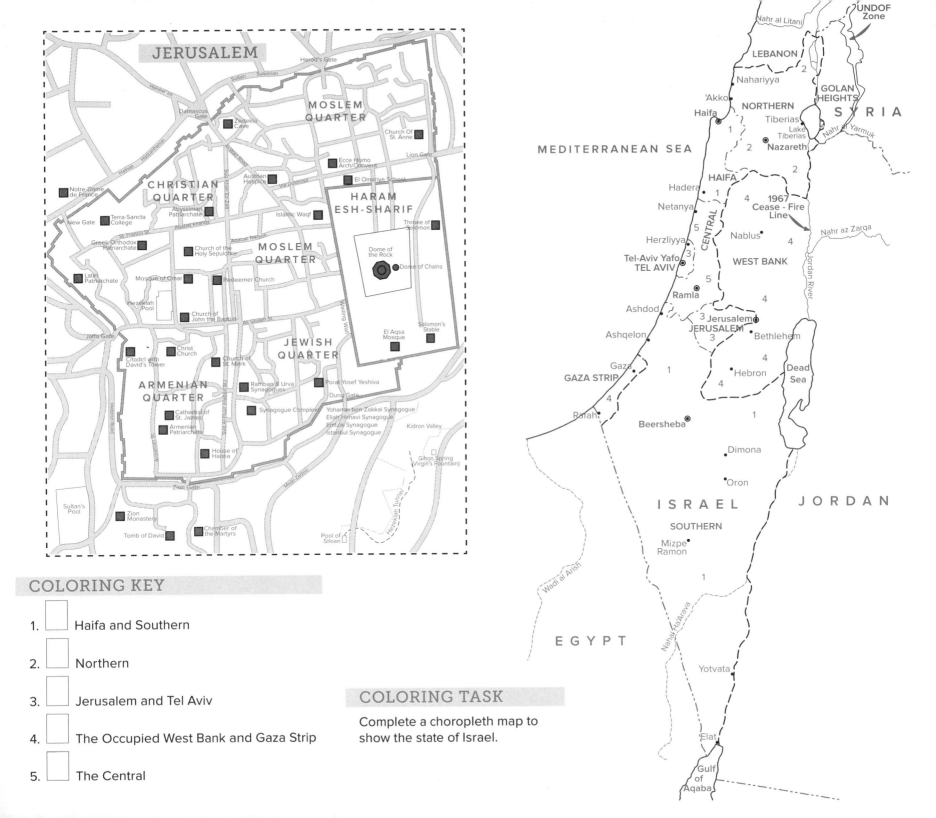

JERUSALEM

MOSLEM QUARTER

CHRISTIAN QUARTER

HARAM ESH-SHARIF

MOSLEM QUARTER

Dome of the Rock

Dome of Chains

JEWISH QUARTER

ARMENIAN QUARTER

COLORING KEY

1. Haifa and Southern

2. Northern

3. Jerusalem and Tel Aviv

4. The Occupied West Bank and Gaza Strip

5. The Central

COLORING TASK

Complete a choropleth map to show the state of Israel.

Afghanistan

AFGHANISTAN WAS INVADED BY THE UNITED STATES AND THE UNITED KINGDOM on October 7, 2001 as part of Operation Enduring Freedom (OEF) in response to the September 11, 2001 attacks against the U.S.

Afghanistan is a landlocked and mountainous country that has suffered from instability and conflict during its modern history. Much of the country's infrastructure is poor, its economy is weak, and many of its people are refugees. Afghanistan sits in a strategic position between the Middle East, central Asia, and the Indian subcontinent along the ancient "Silk Route." Despite its rugged terrain and harsh climate, it has been the scene of many conflicts. It was at the center of the so-called "Great Game" in the nineteenth century, when Imperial Russia and the British Empire in India vied for influence.

The Taliban emerged as the dominant group governing Afghanistan in 1996 and took control of about 90 percent of the country. The Taliban were made up of mainly Pashtun Afghans who were Islamic scholars, but they evolved into an extreme version of Islam and were only recognized as the legitimate rulers of Afghanistan by three other countries. After the U.S. embassy attacks in Africa in 1998 and the September 11, 2001 attacks, the Taliban came under increasing world pressure to hand over Osama bin Laden, who was believed to have orchestrated the attacks and to be hiding somewhere in Afghanistan. After the Taliban refused to hand over bin Laden, the U.S. launched air attacks, followed by an invasion, finally removing the Taliban from power.

Since the fall of the Taliban administration in 2001, the Islamic group have reformed and regrouped. It is now a resurgent force, particularly in the south and east, and the Afghan government has struggled to extend its authority beyond the capital and to forge national unity.

10 FACTS ABOUT AFGHANISTAN

1. **Full name:** Islamic Republic of Afghanistan
2. **Population:** 29.1 million (UN, 2010)
3. **Capital and largest city:** Kabul
4. **Area:** 251,773 sq miles (652,225 km²)
5. **Major languages:** Dari, Pashto
6. **Major religion:** Islam
7. **Life expectancy:** 45 years (men), 45 years (women) (UN)
8. **Monetary unit:** 1 Afghani = 100 puls
9. **Main exports:** Fruit and nuts, carpets, wool, opium
10. **GNI per capita:** U.S. $370 (World Bank, 2009)

COLORING TASK

Color the country to show the key
provinces that make up Afghanistan.

Iraq

IRAQ STRADDLES THE TIGRIS AND EUPHRATES RIVERS and stretches from the Gulf to the Anti-Taurus Mountains, covering the land that was once the ancient civilization of Mesopotomia—one of the cradles of human civilization.

The country became rich throughout the 1970s as a result of the huge oil reserves found there, and Saddam Hussein became the leader of the Ba'ath (Renaissance) and president in 1979. A long, protracted war with Iran began in September 1980 and continued until August 1988 (it was the longest conventional war of the twentieth century), seeing many casualties. It came to an end when Iran accepted a UN-backed peace plan after major setbacks. The 1990s saw Iraq invade neighboring Kuwait, which sparked the Gulf War in which the U.S.-led coalition expelled Iraqi forces from Kuwait. Major international sanctions followed, having a devastating effect on Iraqi society and the economy.

Operation Iraqi Freedom, or the Iraq War, began on March 20, 2003. The U.S. led a multinational coalition of troops in order to remove Saddam Hussein from power, believing that the Iraqi president was developing weapons of mass destruction (WMD). After three weeks of fighting, the coalition entered Baghdad and the Iraqi leader's grip on power there had withered. While many Iraqis appeared jubilant at his removal, several insurgent groups took to random bomb attacks, and the country descended into sectarian warfare. The newly formed government struggled to gain control of Iraq until a "surge" of U.S. troops in late 2007 began to push out of the cities and provinces they had long contested.

The country remains volatile, and disputes with the autonomous Kurdistan region over the oil-rich city of Kirkuk have threatened to derail progress toward political stability.

10 FACTS ABOUT IRAQ

1. **Full name:** Republic of Iraq
2. **Population:** 31 million (UN, 2010)
3. **Area:** 169,235 sq miles (438,317 km²)
4. **Capital:** Baghdad
5. **Major languages:** Arabic, Kurdish
6. **Major religion:** Islam
7. **Life expectancy:** 68 years (men), 73 years (women) (UN)
8. **Monetary unit:** Iraqi dinar
9. **Main exports:** Crude oil
10. **GNI per capita:** U.S. $2,210 (World Bank, 2009)

TURKEY

Zakhu •   • Al 'Amadiyah
DAHUK
◎ Dahuk

ARBIL

Tall 'Afar •
  ◎ Al Mawsil (Mosul)
  • Irbil   • Dukart Buhairat

• Al Quwayr

NINAWA

AS SULAYMANIYAH

Baba Gurgur •   ◎ As Sulaymaniyah
  • Chamchamal   • Khurma
  Kirkuk
AT-TA'MIN
  • Tawuq   • Halabjah

SYRIA

  • Tuz Khurmatu

IRAN

  • Kifri
◎ Tikrit   • Qarah Tappah

Euphrates R. • Anah

SALAH AD DIN   • As Sa'Diyah   • Tolafarush
  Balad •   • Al Miqdadiyah
Tharthar   • Al Khalis
Lake   • Al Jadidah   **Ba'Qubah**   • Mandali

• Hit   **DIYALA**

**SYRIAN
DESERT**
  AL ANBAR   • Ar Ramadi   • Al A Zamiyah
  • Al   Fallujah
  • Al Habbaniyah   **BAGDAD** ★ **Baghdad**
  Haur al-Habbaniya   • Sal Man Pak
• Ar Rutbah   • Al Mahmudiyah
IRAQ   • Al Latifiyah   • Al Aziziyah
  • Al Musayyib   • Az Zubaydiyah
  Razzaza Lake   **BABIL**   **WASIT**   • Ali Al Gharbi

JORDAN   • An Nu'Maniyah   • **Al Kut**
  Karbala'   • Haur Dalmg
  • Al Hillah   • Haur as-Sadiya   **MAYSAN**
  • Kut Al Hayy   • Kumayt

  • Al Kufah   **Ad Diwaniyah**   • Qal At Sukkar
KARBALA'   • An Najaf   **AL-QADISIYAH**   • Qal At Salih   ◎ **Al Amarah**
  • Al Jaarah   • Qawam Al Hamzah   • Ar Rifa   • Musay Idah
  • Al Kharm   • Qal At Salih
  • Ash Shinafiyah   • Ar Rumaythah • Al Majarr Al Kabir
  Euphrates R. • Ash Shatrah   • As Sulayb

  • As Samawah   • Al Khidr   **DHI QAR**   • Ash Shanin
AN NAJAF   • An Nasiriyah ◎   • Duwa
  Suq Ash Shuyukh
  Haur al-Hammar
AL MUTHANNA   • **Al Basrah**
  • Az Zubayr
AL-BASRAH

**PERSIAN
GULF**

KUWAIT

SAUDI ARABIA

COLORING TASK

Color the country to show the
key states that make up Iraq.

Global Distribution of Internet Access

THE INTERNET IS A GLOBAL SYSTEM OF INTERCONNECTED COMPUTER NETWORKS. The template for the Internet began in October 1969, with a system called ARPANET that connected computers at the research center at the University of California in Los Angeles to the Stanford Research Institute. By 1983, ARPANET was replaced with the TCP/IP protocol, which became the most widely used network protocol in the world. By 1990, universities in North America had become connected to research facilities in Europe and the launch of the World Wide Web followed, allowing computers to connect to each other through telecommunications cables.

From 2011 onward, the Internet is expected to grow significantly in Brazil, Russia, India, and China. These countries have large populations and moderate to high economic growth, but still low Internet penetration rates. In 2009, they represented about 45 percent of the world's population and had approximately 600 million Internet users, but by 2015, the number of Internet users in these countries is expected to double.

Internet Submarine Cables

In January 2008, tens of millions of people in the Middle East and Asia were left without access to the Internet after an undersea cable was accidentally cut. The cable that runs under the Mediterranean Sea from Italy to Egypt is part of the longest undersea cable in the world. It is 24,500 miles (39,428 km) long and stretches from Germany, across the Middle East, to Australia and Japan. The result of the break in the cable meant that many Internet users in Saudi Arabia, Egypt, and India struggled to get any Internet access. As many as 70 percent of Internet users in Egypt were unable to get online, while in India there was a 50 percent cut in bandwidth which caused major problems for many of the rapidly expanding high-tech companies. It took several days to fix the cable before Internet was finally restored.

TOP 10 COUNTRIES WITH THE HIGHEST NUMBER OF INTERNET USERS

(2010)

Country	Population	Internet users
1. China	1,330,141,295	420,000,000
2. United States	310,232,863	239,893,600
3. Japan	126,804,433	99,143,700
4. India	1,173,108,018	81,000,000
5. Brazil	201,103,330	75,943,600
6. Germany	82,282,988	65,123,800
7. Russia	139,390,205	59,700,000
8. United Kingdom	62,348,447	51,442,100
9. France	64,768,389	44,625,300
10. Nigeria	152,217,341	43,982,200

Source: Internet World Stats

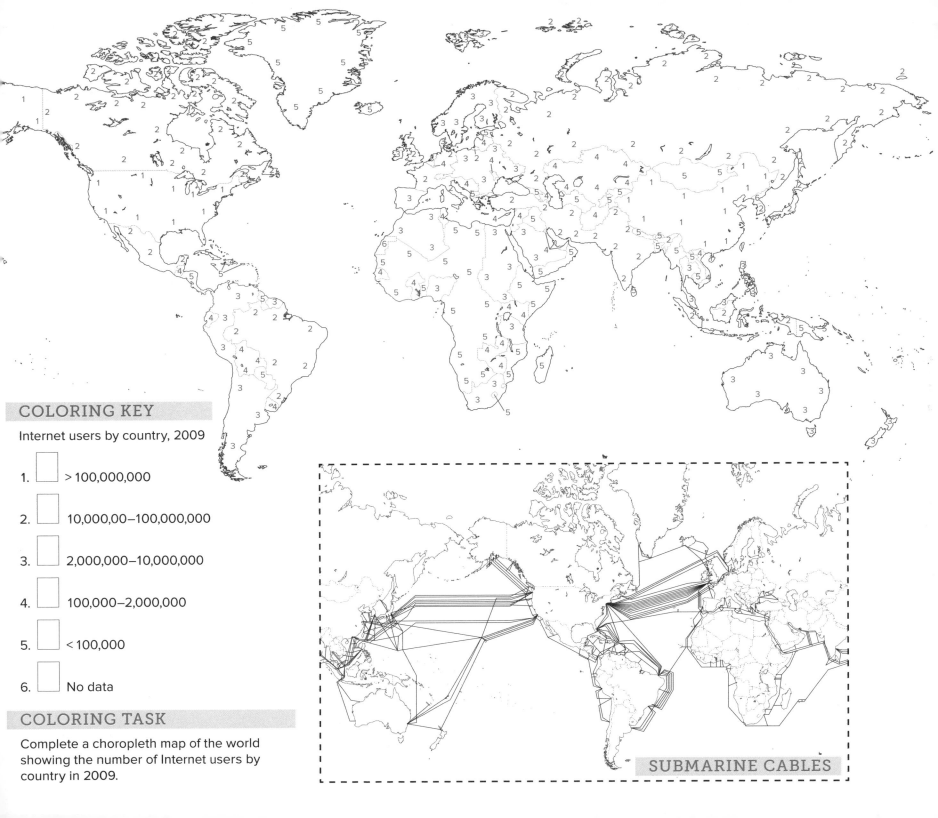

COLORING KEY

Internet users by country, 2009

1. ☐ > 100,000,000

2. ☐ 10,000,00–100,000,000

3. ☐ 2,000,000–10,000,000

4. ☐ 100,000–2,000,000

5. ☐ < 100,000

6. ☐ No data

COLORING TASK

Complete a choropleth map of the world showing the number of Internet users by country in 2009.

SUBMARINE CABLES

Global Distribution of Cell Phones

THERE ARE MORE THAN 5 BILLION CELL PHONE CONNECTIONS worldwide, and more than a billion of these were added during 2009 and 2010. The 4 billion connections mark was passed at the end of 2008, and it is predicted that there will be 6 billion connections worldwide by the middle of 2012. That's almost one cell phone for every person on the planet!

In many countries penetration is now more than 100 percent, and there is more than one connection per person in the country. This is in sharp contrast to when the cell phone was launched in 1987, and the industry predicted a maximum of 10,000 phones. It has now become the most prolific consumer device on the planet. It is in India and China where the main growth exists, accounting for 47 percent of global cell phone connections at the end of June 2010.

The world's largest individual cell operator is China Mobile, with over 500 million cell phone subscribers.

Technological Leapfrogging: Afghanistan

The ongoing war in Afghanistan since 2001 has left the country littered with land-mines, and laying telephone cables is an extremely dangerous business that few companies wish to attempt. Any attempt to put new wires in the ground would take decades and be prohibitively expensive. Therefore, the country's largest telephone operator, Roshan, has "leapfrogged" the need for a fixed line telephone infrastructure and moved into the much safer wireless cell sector. In 2008, Roshan had nearly 2 million cell phone subscribers in Afghanistan, making up 43 percent of the total market. Seventy-two percent of the Afghan population is covered by a cell phone signal, and this is compared to only one person in 100 having a fixed telephone line.

TOP 10 COUNTRIES WITH THE HIGHEST NUMBER OF CELL PHONE USERS

(2007)

1.	China	547,286,000
2.	European Union	466,000,000
3.	India	362,300,000
4.	United States	255,000,000
5.	Russia	170,000,000
6.	Brazil	120,980,000
7.	Japan	107,339,000
8.	Germany	97,151,000
9.	Pakistan	88,020,000
10.	Indonesia	81,835,000

Source: CIA World Factbook

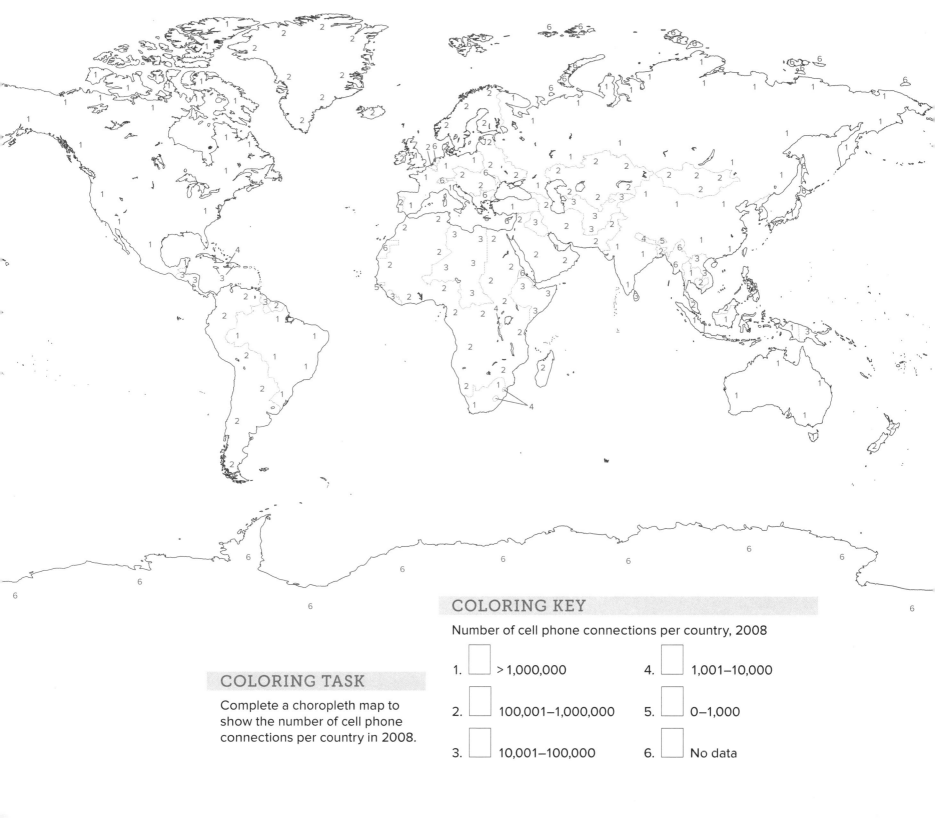

COLORING TASK

Complete a choropleth map to show the number of cell phone connections per country in 2008.

COLORING KEY

Number of cell phone connections per country, 2008

1. ☐ > 1,000,000

2. ☐ 100,001–1,000,000

3. ☐ 10,001–100,000

4. ☐ 1,001–10,000

5. ☐ 0–1,000

6. ☐ No data

MAPS IN FULL COLOR

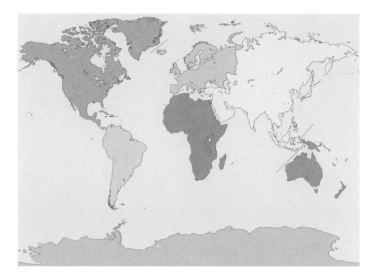

Page 11, Continents and Oceans

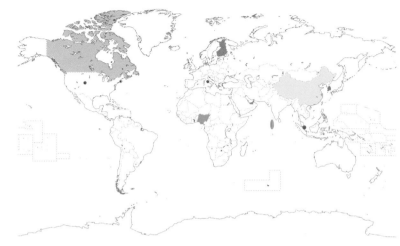

Page 13, Political—Countries of the World

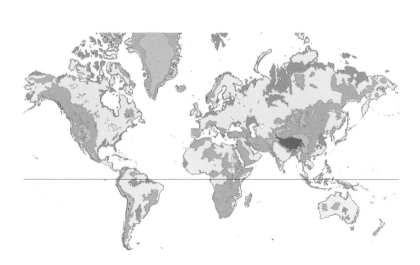

Page 15, Physical Geography

Page 17, North America—Political and Physical

Page 19, South America—Political and Physical

Page 21, Europe—Political and Physical

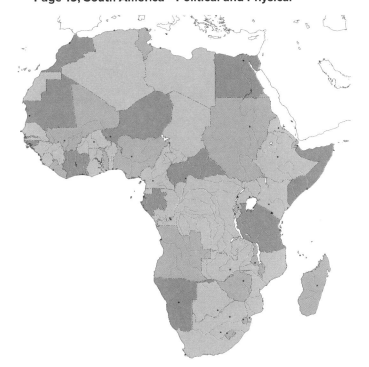

Page 23, Africa—Political and Physical

Page 25, Asia—Political and Physical

COLOR YOURSELF **SMART** GEOGRAPHY MAPS IN FULL COLOR

Page 27, Australia and Oceania—Political and Physical

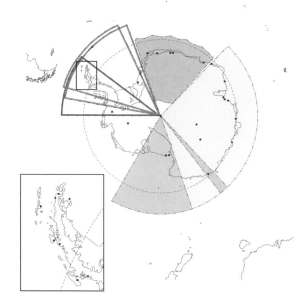

Page 29, Antarctica—Political and Physical

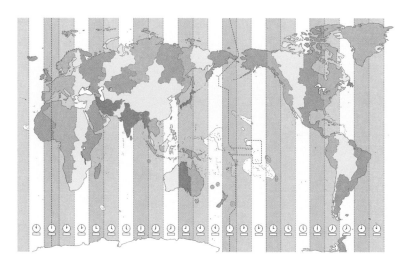

Page 31, World Time Zones

Page 33, The United States

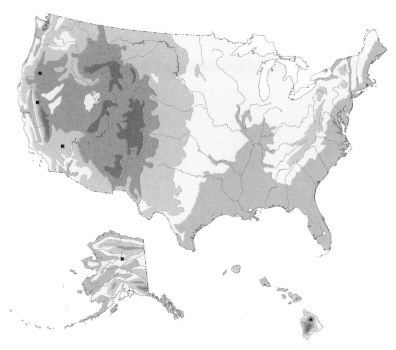

Page 35, Physical Features of the United States

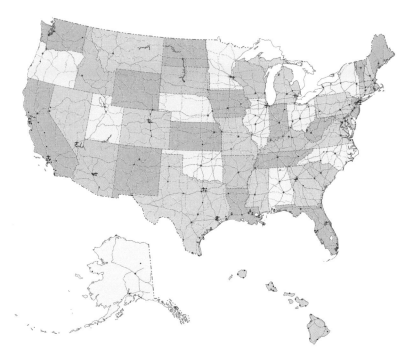

Page 37, Political Map of the United States

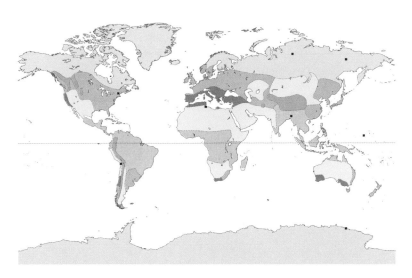

Page 39, World Climate Zones

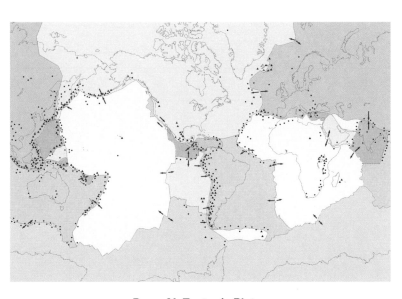

Page 41, Tectonic Plates

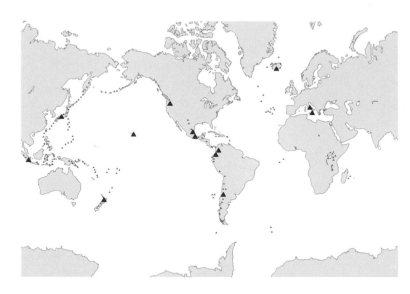

Page 43, Features of a Volcano

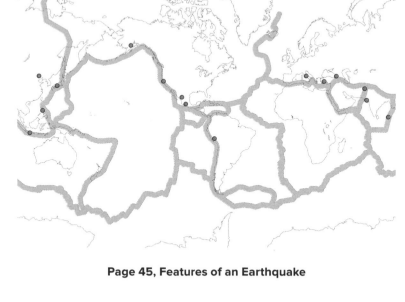

Page 45, Features of an Earthquake

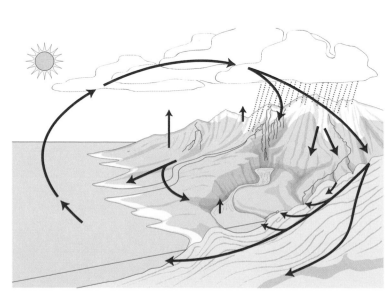

Page 47, The Hydrological Cycle

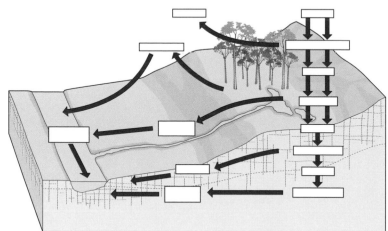

Page 49, The River Basin System

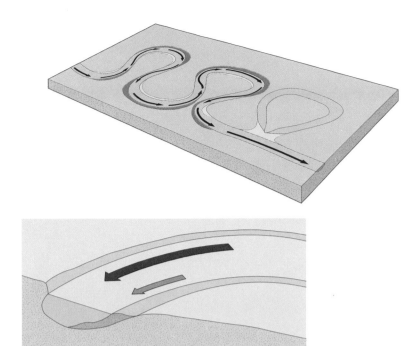

Page 51, River Erosion and Deposition

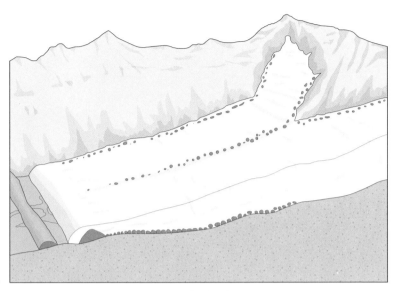

Page 53, Glaciers

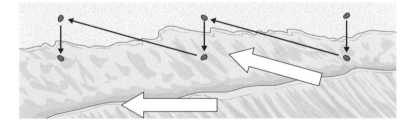

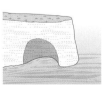

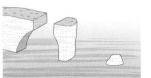

Page 55, Coastal Processes

COLOR YOURSELF **SMART**
GEOGRAPHY MAPS IN FULL COLOR

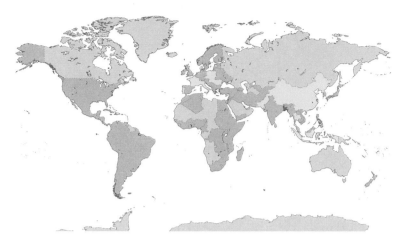

Page 57, Population Density

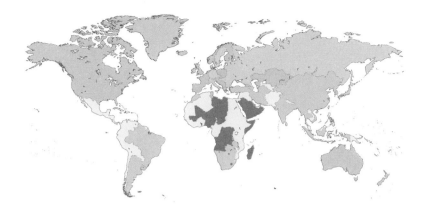

Page 59, Population Change

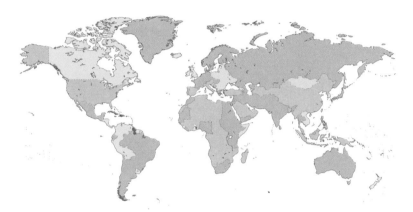

Page 61, Migration to the United States

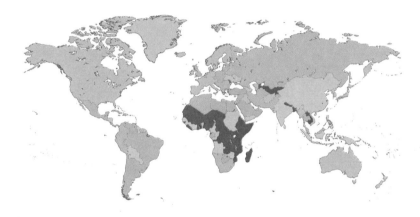

Page 63, Wealth—GDP per Capita

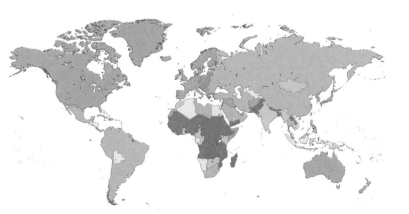

Page 65, Quality of Life

Page 67, The Wretched Dollar

COLOR YOURSELF **SMART**
GEOGRAPHY MAPS IN FULL COLOR

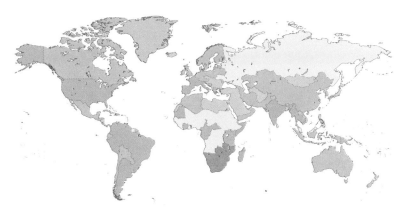

Page 69, HIV Infection Rates

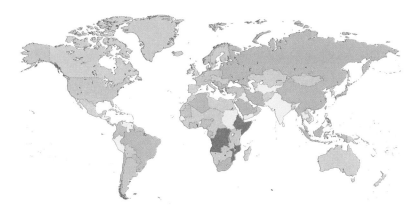

Page 71, How Hungry Are You?

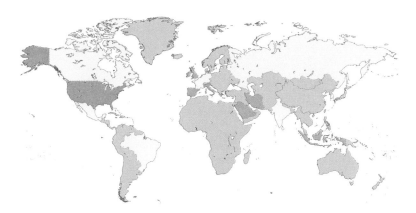

Page 73, Known Oil Reserves

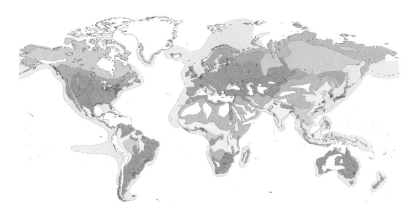

Page 75, Food Production

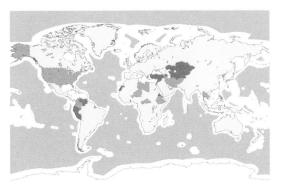

Page 77, The Law of the Sea Treaty

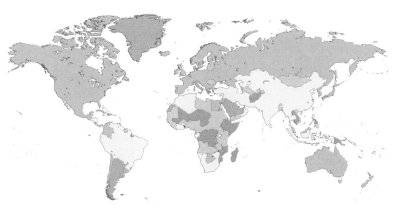

Page 79, Population without Access to Safe Water

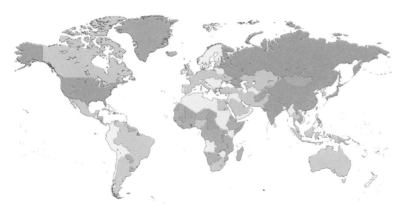

Page 81, The Enhanced Greenhouse Effect

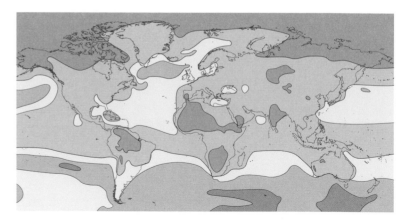

Page 83, Effects of Climate Change

Page 85, Ecological Footprints of Countries

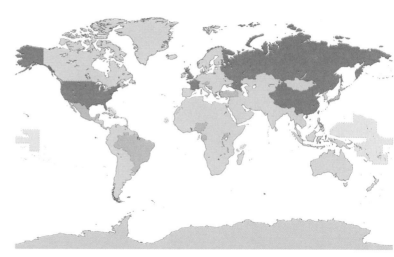

Page 87, Membership of the United Nations

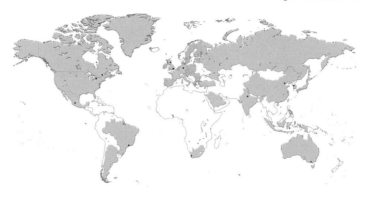

Page 89, The G20

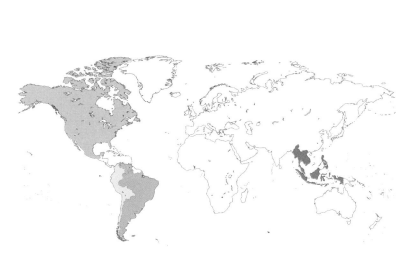

Page 91, Trading Blocs

Page 93, The European Union

Page 95, Brazil

Page 97, Russia

COLOR YOURSELF **SMART**
GEOGRAPHY MAPS IN FULL COLOR

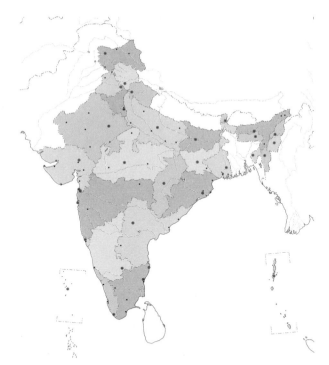

Page 99, India

Page 101, China

Page 103, Countries at War

Page 105, Israel and the Middle East

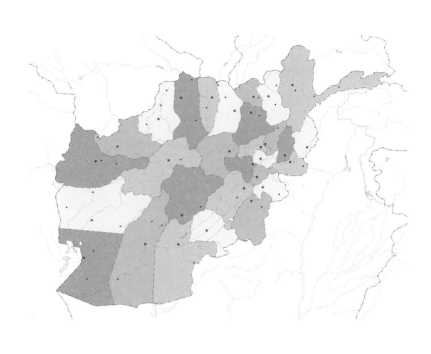

Page 107, Afghanistan

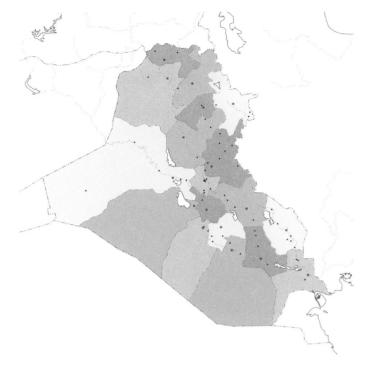

Page 109, Iraq

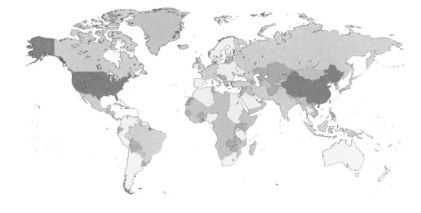

Page 111, Global Distribution of Internet Access

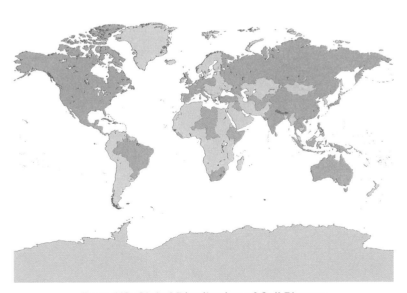

Page 113, Global Distribution of Cell Phones

QUIZ

1. How much of Earth's surface is actually "earth?"
   a) 71 percent
   b) 100 percent
   c) 29 percent
   d) 53 percent

2. Which country in the world has the highest density of millionaires?
   a) France
   b) Singapore
   c) U.S.
   d) Japan

3. What is the longest river in the world?
   a) Amazon
   b) Euphrates
   c) Rhine
   d) Nile

4. What is the largest city in North America?
   a) Mexico City
   b) New York City
   c) Los Angeles
   d) Toronto

5. How many countries are there in South America?
   a) 5
   b) 12
   c) 9
   d) 15

6. What is the longest river in Europe?
   a) River Thames
   b) Rover Po
   c) River Volga
   d) River Rhone

7. Approximately how many people live in Africa?
   a) 2 billion
   b) 500 million
   c) 300 million
   d) 1 billion

8. Asia's three dominant financial centers are . . . ?
   a) Mumbai, Hong Kong, Shanghai
   b) Tokyo, Mumbai, Lahore
   c) Tokyo, Hong Kong, Singapore
   d) Delhi, Singapore, Dhaka

9. Approximately how many people live in Australia and Oceania?
   a) 14 million
   b) 21 million
   c) 17 million
   d) 25 million

10. How much of the land in Antarctica is not covered in ice?
    a) 1 percent
    b) 2.4 percent
    c) 3 percent
    d) 2 percent

11. Greenwich meantime has the same time as London time during winter time, but how much ahead of GMT is London during summer time?
    a) 1 hour
    b) 30 minutes
    c) 2 hours
    d) 12 hours

12. New York state is home to how many species of wild orchids?
    a) 12
    b) 34
    c) 58
    d) 97

13. The U.S. covers approximately how many million square miles?
    a) 3.8
    b) 2.4
    c) 1.6
    d) 4.3

14. The population of the U.S. is estimated at?
    a) 410,678,972
    b) 310,232,863
    c) 298,877,892
    d) 103,567,981

15. What was the highest recorded temperature in the world?
    a) 113°F (45°C)
    b) 120°F (49°C)
    c) 142°F (61°C)
    d) 136°F (58°C)

16. How many people die, on average, as a result of earthquakes each year?
    a) 1,000
    b) 10,000
    c) 57
    d) 100,000

17. How many "active" volcanoes are there thought to be in the world?
    a) 1,510
    b) 1,067
    c) 34
    d) 876

18. How powerful, on the Richter scale, was the 1989 San Francisco earthquake?
    a) 7.1
    b) 6.9
    c) 6.5
    d) 4.2

19. How much of all the world's water is contained in the seas and oceans as salt water and is unsuitable for use by humans, animals, and terrestrial plants?
    a) 54 percent
    b) 67 percent
    c) 86 percent
    d) 97 percent

20. Which two rivers both begin in Switzerland?
    a) Thames and Tee
    b) Rhine and Volga
    c) Po and Rhone
    d) Rhine and Rhone

21. What is the land that gets flooded when a river overflows called?
    a) Wet perimeter
    b) Flood plain
    c) Watershed
    d) Water break

22. What type of moraine is left behind at the maximum advance of a glacier?
    a) Terminal
    b) Recessional
    c) Lateral
    d) Medial

23. What type of waves form if the shore is shallow and the wave spills forward for a long distance?
    a) Destructive waves
    b) Concordant waves
    c) Constructive waves
    d) Discordant waves

24. The United Nations Population Division expects the world population to grow from 6 billion to what by 2050?
    a) 7.8 billion
    b) 9.3 billion
    c) 12.6 billion
    d) 11.2 billion

25. Yemen has a high fertility rate of . . . ?
    a) 6.7 children per woman
    b) 4.5 children per woman
    c) 8.9 children per woman
    d) 7.2 children per woman

26. The U.S. has around 37 million foreign-born residents, making up what percentage of the total population?
a) 2.3 percent
b) 25 percent
c) 45 percent
d) 12.4 percent

27. How much of the of the world's wealth is found in North America and western Europe?
a) 34 percent
b) 46 percent
c) 11 percent
d) 87 percent

28. Which country, with a score of 0.939, tops the UN HDI list in 2010?
a) Sweden
b) Luxembourg
c) Norway
d) U.S.

29. Section A of MDG 1 is to halve the proportion of people who live on the equivalent of U.S. $1 a day or less. In 2002, what percentage of the world population lived on this amount?
a) 17 percent
b) 12 percent
c) 7 percent
d) 2 percent

30. Which of these is not a way in which HIV can be transmitted?
a) Having sexual intercourse with an infected partner
b) Injecting drugs using a needle or syringe that has been used by someone who is infected
c) A mother passing the infection to her baby during pregnancy, labor, delivery, or breastfeeding
d) Shaking hands with an infected person

31. The UN Food and Agricultural Commission (FAO) estimates this number of people suffered from undernutrition worldwide in 2009?
a) 2.23 billion
b) 100 million
c) 567 million
d) 1.02 billion

32. The output of oil has risen significantly from less than a million barrels a day in 1900 to around how many million barrels today?
a) 83
b) 24
c) 56
d) 77

33. A typical farm in the U.S. in 1830 took about 300 hours of work to produce 100 bushels of wheat. By 2010 the process takes how many hours to produce the same amount of wheat?
a) 200
b) 3
c) 60
d) 24

34. The Law of the Sea Treaty is formally known as the Third United Nations Convention on the Law of the Sea and is often referred to as what?
a) LST
b) UNCLOS III
c) LSEA3
d) UNSEATR II

35. Around how much of the world's population experience water shortages?
a) 34 percent
b) 12 percent
c) 40 percent
d) 6 percent

36. Which of the following is evidence that global warming is taking place?
a) Global temperatures have increased since 1850
b) The Arctic ice sheet is melting faster than at any time since records began
c) Frozen peat bogs in Siberia are thawing, releasing billions of metric tons of methane gas into the atmosphere
d) The amount of snow melting in Greenland each summer has increased by 30 percent since 1975

37. How many years do scientists think it will take for the Greenland ice sheet to melt completely?
a) 1,000
b) 780
c) 500
d) 300

38. The term "ecological footprint" refers to . . . ?
a) The size of your garbage can
b) A calculation to estimate the area of Earth's land and water that is needed to supply resources to an individual or a group of people
c) Your shoe size
d) A measure of how much food you consume

39. There are currently how many member states of the United Nations?
a) 192
b) 204
c) 156
d) 197

40. Which of the following countries is not a member of the G20?
a) Brazil
b) China
c) Italy
d) Spain

41. OPEC stands for:
a) Oil and Petrol Economic Countries
b) Organization of Petroleum Exporting Countries
c) Oil and Plastic Exporting Countries
d) Oil, Plastic, Etc., Producers

42. The European Union (EU) is an economic and political union of how many countries located in Europe?
a) 34
b) 12
c) 27
d) 22

43. Brazil is home to over half of South America's population, but where does it come in the top 10 most populated countries in the world?
a) Third
b) Sixth
c) Ninth
d) Fourth

44. Russia has the world's largest reserves of natural gas and is the second-largest producer of oil after Saudi Arabia. These two natural resources accounted for how much of Russia's exports in 2007?
a) 56 percent
b) 76 percent
c) 60 percent
d) 46 percent

45. India is today the fourth-largest economy in the world and is a leading world exporter with an annual growth rate of how much?
a) 8 percent
b) 4 percent
c) 7 percent
d) 2 percent

46. The Chinese steel industry produces one-third of the world's output each year, but how many million metric tons is this?
a) 234
b) 600
c) 780
d) 500

47. The former Socialist Federal Republic of Yugoslavia began to break up in 1991. Eventually, after four years of conflict, a peace agreement was signed where in November 1995?
a) London, England
b) Dayton, Ohio
c) Moscow, Russia
d) Paris, France

48. The GNI per capita of Israel is U.S. $25,740, while the GNI per capita of Palestine is . . . ?
a) U.S. $1,230
b) U.S. $12,465
c) U.S. $23,780
d) U.S. $9,870

49. The October 2001 invasion of Afghanistan was known as what?
a) The End of an Era
b) Game Over
c) Operation Enduring Freedom
d) Call of Duty

50. The capital city of Iraq is what?
a) Kabul
b) Baghdad
c) Tehran
d) Lima

51. In 2009, Brazil, Russia, India, and China represented about 45 percent of the world's population and had approximately how many million Internet users?
a) 600
b) 500
c) 900
d) 700

52. There are more than 5 billion cell phone connections worldwide, but how many of these were added during 2009 and 2010?
a) 400 million
b) 1 billion
c) 700 million
d) 2 billion

QUIZ ANSWERS

1.	c	27.	b
2.	b	28.	c
3.	d	29.	a
4.	a	30.	d
5.	b	31.	d
6.	c	32.	a
7.	d	33.	b
8.	c	34.	b
9.	b	35.	c
10.	d	36.	all
11.	a	37.	d
12.	c	38.	b
13.	a	39.	a
14.	b	40.	d
15.	d	41.	b
16.	b	42.	c
17.	a	43.	b
18.	b	44.	c
19.	d	45.	a
20.	d	46.	d
21.	b	47.	b
22.	a	48.	a
23.	c	49.	c
24.	b	50.	b
25.	a	51.	a
26.	d	52.	b